VOLUME 19
**American Night Fighters Pacific Theatre
1943–1945**

MICHAEL JOHN CLARINGBOULD

Avonmore Books

Pacific Profiles Volume 19
American Night Fighters Pacific Theatre 1943–1945

Michael John Claringbould

ISBN: 9781764193702

First published 2025 by Avonmore Books
Avonmore Books
PO Box 217
Kent Town
South Australia 5071
Australia

Phone: (61 8) 8431 9780
avonmorebooks.com.au

A catalogue record for this book is available from the National Library of Australia

Cover design & layout by Diane Bricknell

© 2025 Avonmore Books.

No part of this book may be reproduced or transmitted in any form or by any means, electronic or mechanical, including photocopying or recording, or by any information storage and retrieval system, without permission in writing from the publisher.

Front cover: The varied schemes which appeared on Pacific night fighters are sampled here (top to bottom): an F4U-2 Corsair of VF(N)-532, Douglas P-70 Gloria from Detachment A of the 6th NFS, P-38J squadron number 305 of the 419th NFS, P-61A Midnight Mickey of the 6th NFS and a P-61B of the 550th NFS, squadron number 330 Night Vision. These appear as Profiles 83, 5, 28, 7 and 67.

Back cover: The commander of the 6th NFS Detachment A, Captain Warren Hornaday, climbs at dusk above the overcast overhead Port Moresby's Hombrom's Bluff where a ground radar operator has vectored him towards an incoming bogey. The colourful motif of his P-70A-2 is illustrated in Profile 4.

Contents

About the Author 5

Glossary & Abbreviations 6

Map 7

Introduction 8

Chapter 1   Technical Notes 11

Chapter 2   6th Night Fighter Squadron Detachments A & B 19

Chapter 3   6th NFS (Reformed) 25

Chapter 4   418th Night Fighter Squadron 31

Chapter 5   419th Night Fighter Squadron 39

Chapter 6   421st Night Fighter Squadron 47

Chapter 7   547th Night Fighter Squadron 55

Chapter 8   548th Night Fighter Squadron 59

Chapter 9   549th Night Fighter Squadron 65

Chapter 10   550th Night Fighter Squadron 69

Chapter 11   VF(N)-75 & VF(N)-101 75

Chapter 12   VMF(N)-531 79

Chapter 13   VMF(N)-532 87

Chapter 14   Miscellaneous 91

Sources & Acknowledgments 94

Index of Names 95

Michael Claringbould at a paraglider launch not far from Mount Blanc, France, in September 2024.

About the Author

Michael spent his formative years in Papua New Guinea in the 1960s, during which he became fascinated by the many WWII aircraft wrecks which still lie around the country.

Michael has served widely overseas as an Australian diplomat, including in South East Asia and throughout the South Pacific where he had the fortune to return to Papua New Guinea for three years commencing in 2003.

Michael has authored and illustrated many books on Pacific War aviation, including over 30 titles for Avonmore Books. His history of the Tainan Naval Air Group in New Guinea, *Eagles of the Southern Sky*, received worldwide acclaim as the first English-language history of a Japanese fighter unit, and was translated into Japanese. An executive member of Pacific Air War History Associates, Michael holds a pilot license and PG4 paraglider rating. He continues to develop his skills as a digital 3D aviation artist, using 3DS MAX, Vray and Photoshop to attain markings accuracy.

Other volumes in this series:

Pacific Profiles Volume One Japanese Army Fighters New Guinea & the Solomons 1942-1944 (2020)

Pacific Profiles Volume Two Japanese Army Bomber & Other Units, New Guinea and the Solomons 1942-44 (2020)

Pacific Profiles Volume Three Allied Medium Bombers, A20 Series, South West Pacific 1942-44 (2020)

Pacific Profiles Volume Four Allied Fighters: Vought F4U Corsair Series Solomons Theatre 1943-1944 (2021)

Pacific Profiles Volume Five Japanese Navy Zero Fighters (land-based) New Guinea and the Solomons 1942-1944 (2021)

Pacific Profiles Volume Six Allied Fighters: Bell P-39 & P-400 Airacobra South & Southwest Pacific 1942-1944 (2022)

Pacific Profiles Volume Seven Allied Transports: Douglas C-47 series South & Southwest Pacific 1942-1945 (2022)

Pacific Profiles Volume Eight IJN Floatplanes in the South Pacific 1942-1944 (2022)

Pacific Profiles Volume Nine Allied Fighters: P-38 series South & Southwest Pacific 1942-1944 (2022)

Pacific Profiles Volume 10: Allied Fighters: P-47D Thunderbolt series Southwest Pacific 1943-1945 (2023)

Pacific Profiles Volume 11: Allied Fighters: USAAF P-40 Warhawk series South and Southwest Pacific 1942-1945 (2023)

Pacific Profiles Volume 12: Allied Fighters: P-51 & F-6 Mustang series New Guinea and the Philippines 1944-1945 (2023)

Pacific Profiles Volume 13: IJN Bombers, Transports, Flying Boats & Miscellaneous Types South Pacific 1942-1944 (2024)

Pacific Profiles Volume 14: Allied Bombers: B-25 Mitchell series Australia, New Guinea and the Solomons 1942-1945 (2024)

Pacific Profiles Volume 15: Allied Bombers B-26 Marauder series Australia, New Guinea and the Solomons 1942-1944 (2024)

Pacific Profiles Volume 16: Allied Bombers B-17 Flying Fortress series Australia, New Guinea and the Solomons 1942-1944 (2024)

Pacific Profiles Volume 17: Allied Air Power Netherlands East Indies 1940-1942 (2025)

Pacific Profiles Volume 18: Allied Bombers RAAF Beauforts and Beaufighters 1942-1945 (2025)

Glossary & Abbreviations

Note: Japanese terms are in italics. Japanese names are presented with the surname first as per Japanese convention.

AI	Airborne Intercept (radar)
Bu Aer	Bureau of Aeronautics Number
CAP	Combat Air Patrol
COMAIRSOLS	Commander, Aircraft, Solomons
COMSOPAC	Commander, South Pacific Area
CRTC	Combat Replacement Training Center
FPO1c	Flying Petty Officer First Class
FG	Fighter Group
FS	Fighter Squadron
GCI	Ground Controlled Interception
IJN	Imperial Japanese Navy
Kokutai	A Japanese naval air group
MAG	Marine Aircraft Group
NFS	Night Fighter Squadron
PT	Patrol Torpedo (boat)
RAAF	Royal Australian Air Force
RAF	Royal Air Force
US	United States
USS	United States Ship
USAAC	United States Army Air Corps
USAAF	United States Army Air Force
USMC	United States Marine Corps
USN	United States Navy
WWII	World War Two

Map

US night fighters operated across the breadth of the entire Pacific theatre. Initial modest deployments began in the Solomons and New Guinea in early 1943. Then in the final two years of war a growing number of squadrons operated throughout the Philippines and the Central Pacific. During the advance to Japan bases at Saipan, Iwo Jima and Ie Shima were utilised, and after the end of the war in August 1945 some squadrons deployed to Japan itself.

Introduction

Welcome to the diverse world of Pacific War night fighters. This volume focuses on their markings and insignia, mostly under-represented to date with the exception of the P-61 Black Widow series. In addition to the P-61, other types illustrated include the P-70 series Nighthawk, the P-38 Lightning, the B-25 Mitchell, the F4U-2 Corsair and the Lockheed PV-1 Ventura. The F6F Hellcat also served as a dedicated night fighter, however, these were carrier-based and thus are not represented in this volume. Neither are the TBM Avengers of VT(N)-90 or numerous PBY squadrons which often performed as night bombers. Three other miscellaneous types associated with the night fighter program are illustrated in the final chapter.

As early as October 1940 the USAAC held discussions at Wright Field in Ohio with Northrop Corporation's chief of research, Vladimir Pavlecka, during which he was given a specification wish-list for an aircraft project titled "Night Interceptor Pursuit Airplane". Then headquartered in Hawthorne, California, the Northrop Corporation had been running for a year as a subcontractor to major aircraft manufacturers. The challenge for the USAAC in obtaining a dedicated night fighter was that the large aircraft manufacturing companies such as Lockheed, Douglas and Boeing were fully committed to other projects. Hence Pavlecka was asked to design an interceptor which would incorporate electronic devices to locate other aircraft in complete darkness.

In these circumstances the American night fighter program took time, with the first production Northrop P-61 not completed until October 1943. For the Pacific the deployment of night fighters was further delayed as the European theatre had priority for aircraft deliveries. In response a series of interim programs was initiated, the first for the USAAC being the P-70 program, a hybrid concept developed from the Douglas A-20A airframe. Experimental programs with the type commenced in April 1942.

Meanwhile those in the frontlines had grown impatient to see measures to counter persistent Japanese night raids in New Guinea and the Solomons. Admiral Ernest King even requested the British Air Ministry to furnish the USN with night fighters, with the sub-text implying he preferred Beaufighters. The request was unsuccessful but shortly thereafter the Fifth Air Force submitted to Washington on 4 February 1943 that:

> … a night fighter squadron is urgently needed to meet enemy tactics of concentrated area night bombing, which is seriously interfering with our night operations. Accuracy of enemy bombing is improving, and, if unopposed heavy losses by us are to be expected.

As a result of these high-level requests, interim night fighters were earmarked for the Pacific. Two separate batches of P-70s operating with the 481st Night Fighter Operations Training Group at Orlando, Florida, were earmarked for New Guinea and the Solomons, to be operated by the 6th Night Fighter Squadron. These P-70s would quickly develop tactics and procedures for radar-controlled night interceptions, but from the start the type proved sub-standard for the role.

To supplement the P-70's lacklustre performance, Pacific units took the initiative of deploying

P-38s which had a better rate of climb and higher ceiling. Loitering at 30,000 feet above airfields, the P-38s usually had to wait for searchlights to illuminate enemy bombers. However, this reliance on searchlights limited interceptions especially as Pacific weather conditions were seldom clear. To bypass the issue several P-38s were fitted with airborne radar. The concept predictably stumbled because of the excessive workload incurred in single-pilot operation.

Frustrated by having to wait a lot longer for the P-61, the 418th NFS took matters into their own hands and converted a dozen B-25H Mitchells into night fighter configuration at Nadzab in March 1944. These served as an interim platform after discarding its P-70s which wound up with the Combat Replacement Training Center. The Mitchell's 75mm cannon and upper turret were removed, and the aircraft's offensive capacity was increased to fourteen forward-firing machine guns. However, with New Guinea skies swept clear of Japanese aircraft at this juncture, these Mitchells were used for daytime strikes instead. It was not until May 1944 that a Fifth Air Force night fighter squadron took delivery of its first P-61.

Meanwhile, the first of the much-promised Northrop P-61 Black Widows had been delivered to the South Pacific on 3 May 1944. The P-61 was soon showcased widely in the US media as an exemplar of American know-how of the times, which it truly was. The initial publicity surrounding the type continued throughout the war and as a result it all but eclipsed other night fighters operating in the Pacific. These had deliberately been hidden from public view due to the program's secrecy, and they soon disappeared into history.

Meanwhile both the USN and USMC had also entertained night operations in considerable secrecy, after first posting pilots to England to learn RAF night fighter tactics. The USN initiative was codenamed Project Affirm which commenced in April 1942 at Quonset Point, Rhode Island. That same year the USMC received authorisation to establish eight night fighter squadrons each with a dozen airframes, by the first half of 1943. However, as we shall see, similar to their USAAF counterparts, both the USN and USMC ramped up interim programs and types. The USMC, in particular, treated its program in considerable secrecy, even in the field, partially explaining why the history of USMC night fighter units continues to be obscure.

The P-61 eventually saw service across the Pacific with eight USAAF squadrons spread between the Fifth, Seventh and Thirteenth Air Forces. The type's first official kill is credited to 6th NFS P-61A *Moonhappy* (see Profile 8) over Saipan on the evening of 30 June 1944. The Black Widow remained in service for a few more years in the reconnaissance role during the Japanese occupation, and back in the US it became involved in weather and ejector seat test programs.

The later night fighter conquests in the Pacific were decisive, however, herein lies an irony. The enemy threat in both New Guinea and the Solomons evaporated with the Japanese withdrawal of air power from Rabaul in early 1944, almost coinciding with the arrival of the P-61. The new type needed to move closer to the Japanese homeland before it could showcase its prowess which it did and impressively so.

Michael John Claringbould
Canberra, Australia, June 2025

A publicity shot of P-70 serial 39-753 Black Magic serving the 481st Night Fighter Operations Training Group at Orlando around late 1943. This P-70 showcases the type's matt black finish, with semi-gloss black applied to the modified nose section. In the background is experimental Northrop YP-61-NO Black Widow serial 41-18887, undergoing trials with the same unit.

USAAF UNIT ASSIGNMENTS

Fifth Air Force

6th NFS(Detachment A)
418th NFS
421st NFS
547th NFS

Seventh Air Force

6th NFS
548th NFS
549th NFS

Thirteenth Air Force

6th NFS(Detachment B)
419th NFS
550th NFS

Air force assignments of the eight USAAF night fighter squadrons and two forward detachments that feature in this book.

CHAPTER 1
Technical Notes

Northrop P-61 Black Widow series

The P-61's twin-boom tail provided platform stability behind prop wash and spoiler ailerons – termed "spoilerons" – which were developed to assist tight turns. Exhaust flame dampers and gun flash suppressors reduced light coming from the aircraft. The armament pack comprised four 20mm Hispano AN/M2 cannon bolted into the ventral fuselage with 200 rounds per gun. Another four 0.50-inch calibre M2 Browning machine guns were in a remotely operated upper turret, each with 560 rounds per gun. Housed in the fibreglass-covered nose radome was a top-secret Western Electric microwave radar.

The cockpit accommodated a single pilot, with the gunner behind him beneath a bulletproof Plexiglas canopy allowing panoramic visibility. The third crew member was the radar operator who sat in a separate compartment in the rear of the fuselage

In November 1943 the first production P-61A was available for training. Originally camouflaged in Olive Drab and underneath grey, later tests ascertained that overall matt black livery proved visible in moonlight, and experimentation showed that a semi-gloss finish hid the airframe better in night skies. Within six months the first combat-ready P-61As were sent to Pacific night fighter squadrons already operating P-70s and P-38s. The 6th NFS was the first to receive the P-61 in the Pacific.

Northrop dispatched its chief test pilot, John Myers, to the Pacific to assist with the transition process. Skeptical fighter pilots dared the bulky P-61 to mock dogfights and Myers, who once test-flew the Lockheed P-38, would proceed to out-turn combatants, finishing with an aerobatic display. Of this time, he wrote:

> Everywhere we went we were met with raised eyebrows, and there were remarks about it being funny-looking, awkward or odd. They were a surprised bunch before they landed, and P-70 pilots in particular were quick converts. We had instances where crews flew her for the first time in the afternoon, then went out on successful combat missions that night.

The P-61A was the first production version of the type coming off the production line in October 1943. Only the first 37 of the 45 P-61A-1s were fitted with the General Electric remote-control dorsal gun turret, and about half of all P-61As produced left the production line with no turret. This was due to a shortage of turrets which had been commandeered for the B-29 program. The initial turrets with an elongated housing produced buffeting when elevated or rotated. Despite considerable experimentation, the issue was never eliminated, and the field solution was to lock all four 0.50-inch calibre machine guns into the forward horizontal azimuth.

Where the turret was removed the gap was faired over and an extra fuel tank was often fitted.

Such modifications were conducted both in the field and at service squadron level in Hawaii. Those P-61s without a turret complicated the crew situation when the superfluous gunner was left behind on operations. This meant Black Widow missions often carried only a pilot and radar operator, nonetheless, the gunner could also go along as another pair of welcome eyes.

All P-61A-5 models onwards were fitted with upgraded 2,250hp Wright R-2800-65 engines, replacing the initial 2,000hp R-2800-10. Until now all P-61A variants had sported ubiquitous USAAF Olive Drab camouflage, however, the P-61A-5 was the first model to showcase overall satin black which then became a trademark of the type. Twenty P-61A-10s were subsequently modified into P-61A-11 specification, which were fitted with wing hardpoints to carry either two 265 US gallon fuel tanks (later 310 US gallon tanks were fitted), or a variety of ordnance and bombs.

The next substantive version was the P-61B which incorporated a suite of improvements learned from hard-won field experience. The model incorporated improved cockpit heaters, an automatic engine cooling system via cowl activation and oil-cooler ducts with intercoolers. A taxi light was added to the nose landing gear strut. A SCR-720C airborne intercept radar was fitted and the fuselage gondola was extended by eight inches to give more crew comfort. The P-61A's hydraulic maingear doors which sometimes jammed were replaced by link-rod activated doors. These closed the rear doors whenever the maingears were down and locked, preventing debris accumulation in the engine nacelle during take-off and landing. A maingear emergency release would separately lower the undercarriage when there was no hydraulic pressure, a feature which saved many a combat-damaged Black Widow.

Night-vision binoculars with an optical gun sight to enhance targeting were introduced on later P-61B models. These proved so effective that they were retrofitted throughout the fleet including to P-61As. Commencing with the P-51B-5 series, the initial SCR-718 altimeter was replaced with an APN-1 low-altitude altimeter, lending more safety to instrument approaches. An APS-13 tail-warning system was added commencing with the P-61B-10 model along with additional underwing hard points for four 258-gallon drop tanks or an equivalent bomb load. The P-61B-15 model onwards incorporated a modified General Electric Type A-4 turret, then the P-61B-20 model onwards was fitted with the upgraded Type A-7 turret with an improved fire-control system.

Most of these developments were trialled at the night fighter training base at Hammer Field, California, and the Air Proving Command based at Eglin Field, Florida. A final series of developments produced the P-61C, the first of which was accepted by the USAAF in July 1945. However, the Pacific War ended before any saw combat.

TECHNICAL NOTES

P-61A serial blocks:

42-5485 to 5529	P-61A-1 (Olive Drab and grey)
42-5530 to 5564	P-61A-5 (black scheme with white radome)
42-5565 to 5604	P-61A-10 (black scheme hereon)
42-5605 to 5614	P-61A-11
42-5615 to 5634	P-61A-10
42-39348 to 39374	P-61A-10
42-39375 to 39384	P-61A-11
42-39385 to 39397	P-61A-10 and -11

P-61B serial blocks:

42-39398 to 39497	P-61B-1 and -2
42-39548 to 39572	P-61B-10
42-39573 to 39667	P-61B-10, -11, -15, -16 and -25
42-39668 to 39757	P-61B-15
43-8231 to 8236	P-61B-25
43-8237 to 8320	P-61B-20

P-61B 42-39442 on Guam in late 1945 when serving with the 418th NFS. The glossy black finish on the type can be clearly seen.

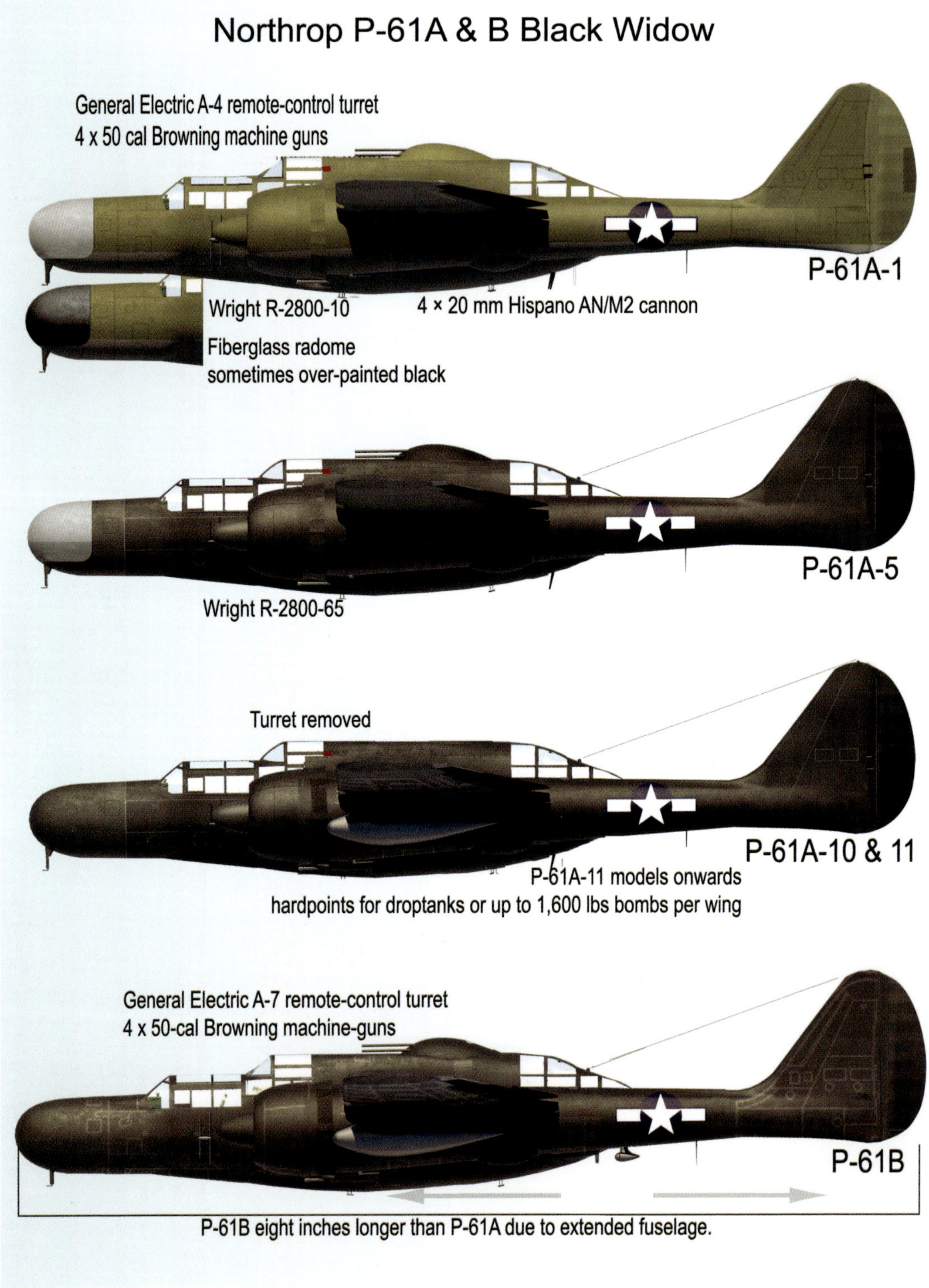

Northrop P-61A & B Black Widow

General Electric A-4 remote-control turret
4 x 50 cal Browning machine guns

4 × 20 mm Hispano AN/M2 cannon

Wright R-2800-10
Fiberglass radome sometimes over-painted black

P-61A-1

Wright R-2800-65

P-61A-5

Turret removed

P-61A-10 & 11

P-61A-11 models onwards hardpoints for droptanks or up to 1,600 lbs bombs per wing

General Electric A-7 remote-control turret
4 x 50-cal Browning machine-guns

P-61B

P-61B eight inches longer than P-61A due to extended fuselage.

This diagram shows technical features of the different P-61A/B models as explained in the text.

TECHNICAL NOTES

Vought F4U-2 Corsair

The USN initiative to produce night fighters was codenamed Project Affirm which commenced in April 1942 at Quonset Point, Rhode Island. The F4U-2 Corsair night fighter first flew on 8 January 1943 and was a modification of the F4U-1 "birdcage" model, with an Airborne Intercept (AI) radar set housed in a bulbous installation on the starboard wing. It thus became the first US single-engine aircraft modified as a night fighter. The antenna was housed in a bulbous radome placed two thirds of the way along the leading edge where the AN/APS-6 radar was mounted. Surprisingly the large bulb minimally affected the Corsair's performance. The outermost port machine gun was removed to restore lateral balance. A six-inch radar scope was added to the centre of the main instrument panel. The USN designation system prescribed an "N" suffix for night fighters, however, the attendant designation F4U-2N was never used.

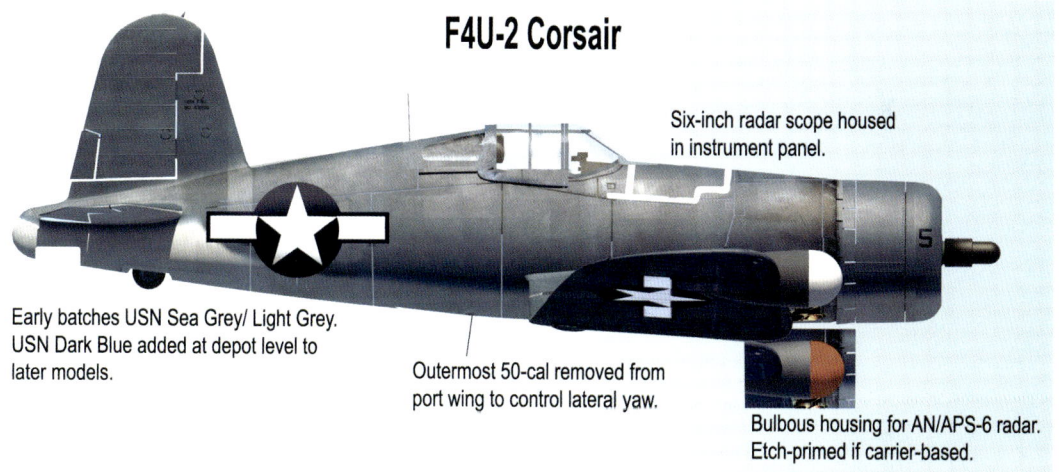

The layout and features of the F4U-2 Corsair, with its bulbous radar housing on the starboard wing.

The three dozen F4U-2s which rolled off the production line (actually 34 in the US and two more converted in the field) were initially camouflaged in the USN Sea Grey and Light Grey scheme. Later batches were repainted in the USN tricolour camouflage with semi-gloss dark blue added to the fuselage and wing topsides at the depot where each painter had a slightly varied technique and hence application. Thus, the night fighter Corsair served in standard USN camouflage schemes, rather than involving matt black schemes as wrongly profiled in many publications.

Although eight USN and two USMC F4U-2 squadrons were deployed to the Pacific, most of the USN squadrons arrived too late to see combat action or were exclusively carrier-based and are thus beyond the scope of this volume. Covered here are one USMC and two USN Corsair night fighter squadrons: VMF(N)-532 and VF(N)-75 that became VF(N)-101.

Douglas P-70 series

Some 59 P-70 night fighter conversions were made using Douglas A-20A airframes. These were armed with four 20mm cannons installed in a ventral pack attached underneath the mid-fuselage. The Perspex "birdcage" was initially retained and painted over, the panels of which were often replaced in the field with aluminium sheeting. The rear flexible defensive twin machine guns were removed and replaced by a radar station. The first P-70 was delivered in April 1942, and the last the following September. The name Nighthawk was officially allocated to the type, however, it was not used in the field.

In the Pacific the P-70A-1 followed the P-70 with 22 modified A-20C airframes delivered to Brisbane in late August 1943, before they were transported to the nearby USAAF Eagle Farm depot on 1 September 1943. These were essentially a P-70 with improved radar equipment installed in the nose. The ventral gun pack was removed, and six 0.50-inch calibre machine guns were installed in the nose. One outlier was serial 42-33147 which was assigned to the 81st Air Depot Group as a hack. Nine more were sidelined in Brisbane and reverted to A-20C configuration to serve with No. 22 Squadron, RAAF. This left the remaining dozen in P-70A-1

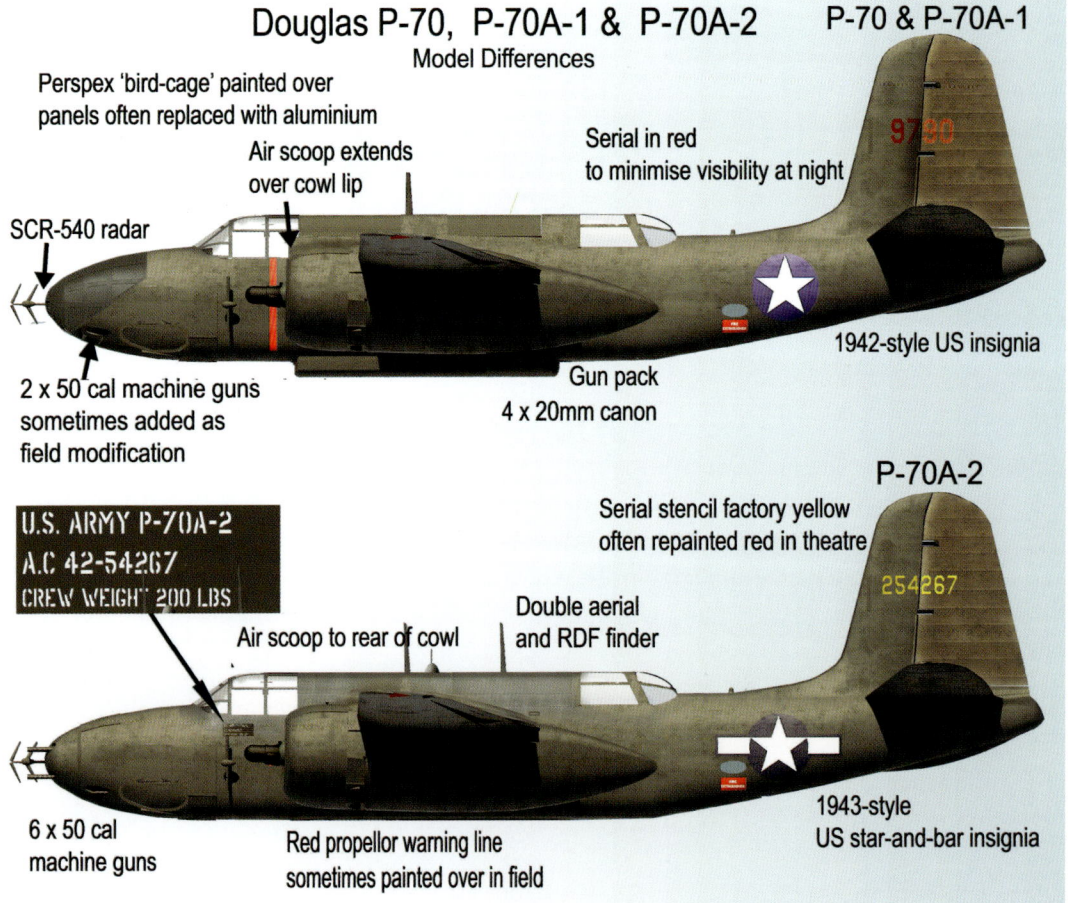

The key technical features of the Douglas P-70, P-70A-1 and P-70-A-2 as explained in the text.

configuration of which a handful had A-20G noses attached for conversion to P-70A-2 status. This was the final P-70 modification to appear in the Pacific which retained the rear canopy, but whose solid nose housed six 0.50-inch calibre machine guns. Sub-variants of the A-1 and A-2 models varied at times due to field modifications. The P-70B was a further development of the series, however none operated in the Pacific.

All P-70s served with at least two squadrons, producing a variety of markings. Serial numbers could be applied in either red or yellow, and the red propeller warning line and/or manufacturer's stencil block could be painted over during repairs or reapplied in the field.

At Guadalcanal the 419th NFS equipped several of their P-70A-1s with SCR-540 radar sets which used a system of four horizontal antennas on each side of the fuselage mounted in the horizontal azimuth. This created a signal strength differential indicating the degree to which the target was offset.

Lockheed PV-1 Ventura

The PV-1 Ventura was built by the Vega Aircraft division of Lockheed (thus the "V" in PV-1). It was modified to USN patrol bomber specifications from the B-34, a derivative of the Lockheed Lodestar. Its fuel capacity of 1,607 US gallons and an ability to carry drop-tanks furnished a healthy range and an ASD-1 search radar was originally fitted. The converted PV-1 night fighter was redesignated PV(N), although the term PV-1 remained interchangeable in USN parlance. Fitted also with a Sperry aerial intercept radar, the operating unit's ground echelon included air and ground control radar operators to direct operations.

Early production runs retained the bombardier's station behind the nose radome, with four side windows and a flat bomb-aiming panel underneath. All fifteen PV-1s assigned to VMF-531(N) were early production models, all of which had served previously in the US and were delivered with worn engines. The front windows were often painted over, and the forward-facing armament comprised six 0.50-inch calibre machine guns in the nose. The dorsal Martin turret held two more 0.50-inch calibre machine guns, whilst two flexible 0.30-inch calibre machine guns were mounted in the rear fuselage.

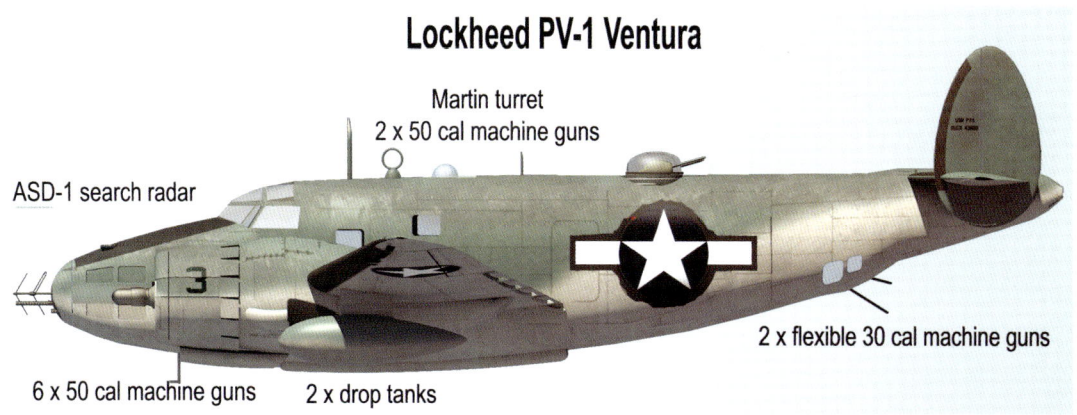

The armament and layout of the Lockheed PV-1 Ventura, as explained in the text.

P-70A-2 serial 42-54268, as depicted in Profile 4, shortly after it was delivered to Port Moresby from Brisbane. Note the red protection covers on the nose guns.

This starboard view of P-70 39-768 Dusty at 12-Mile 'drome, Port Moresby, showcases the aerial system and factory modified forward air scoop over the cowl. The fighter is illustrated in Profile 2. Note the two 20mm cannons mounted in the nose. This was a unique field modification trialled by Detachment A of the 6th NFS.

CHAPTER 2
6th Night Fighter Squadron Detachments A & B

The initial deployment of the 6th NFS Detachments A and B to the Pacific was precipitated by wishing to stop Japanese night harassment attacks, particularly against Guadalcanal. Detachment A served with the Fifth Air Force in New Guinea for eight months, but was recalled to Hawaii in late 1943, whilst Detachment B served with the Thirteenth Air Force on Guadalcanal from where the first P-70 mission was flown on 1 April 1943. Operations were conducted in conjunction with a New Zealand ground radar unit which used the callsign "Kiwi".

Detachment A departed Hawaii on 18 February 1943, with six P-70s and two Liberator LB-30 transports to carry supplies and navigate the long delivery route. One P-70 flown by Second Lieutenant John Meyer disappeared *en route* an hour short of Fiji on 23 February after it disappeared in a weather front near Wallis Island. The remaining five stopped over at Espiritu Santo, where they had their supplementary bomb-bay fuel tanks replaced with an Airborne Intercept (AI) radar. They then proceeded to Port Moresby's Three-Mile 'drome a few days later, staging via Milne Bay, and where they operated under Fifth Fighter Command.

Detachment A's pilots quickly discovered that their P-70s struggled to gain sufficient altitude to intercept their main adversary, G4M1 Bettys launched from Rabaul or Kavieng. Poor climb performance was the major reason why the aircraft failed to perform in the night fighter role. As a result, the unit's pilots lacked enthusiasm to fly the P-70, to the extent they openly discussed how they would rather be flying the 9th FS Lightnings stationed on the other side of the airfield.

The P-70's endurance limitation meant even local patrols were ruled out. When Port Moresby's ground radar picked up inbound bogeys only one or two P-70As were scrambled, which mostly failed to engage the intruders. Furthermore, the P-70s were ferried to nearby Seven Mile 'drome every evening, and then returned early the next morning, a policy introduced by detachment commander Captain Warren Hornaday. The initiative enhanced operating safety as, aside from night bombings, Three-Mile 'drome was surrounded by hills and thus was risky for both night take-offs and landings.

Enemy belligerents planned night raids for moonlight nights to navigate visually, however, New Guinea's weather often conspired against them. Of the 45 IJN sorties launched against Port Moresby in the fortnight commencing 10 May 1943, fifteen were aborted soon after departure, and sixteen more failed to reach Port Moresby due to the weather.

The P-70's AI radar was affected adversely by New Guinea's heat and humidity and even when working accurately interpreting the azimuth and elevation of bogeys required skilled operators. Since P-70 pilots had no radar scope, the radar operators acted as their "eyes". During the intensity of intercepts, the aircraft was therefore only as effective as crew coordination permitted, a collaboration often sorely tested over the interphone. Nonetheless both Detachment A and B scored confirmed kills.

The first success was earned by Detachment B over Guadalcanal in the early hours of 19 April 1943 and was credited to Captain Earl Bennett. After ten No. 705 *Kokutai* G4M1 Bettys departed Rabaul, Bennett intercepted a Betty commanded by Warrant Officer Furuya Sadao at 22,000 feet. This was the service ceiling for a P-70, and it took Bennett's struggling fighter 45 minutes to reach. Whilst Furuya's bomber was captured in Guadalcanal's searchlights for five minutes, Bennett attacked. In a remarkable coincidence, the radar operator on this mission was Technical Sergeant Raymond Mooney, the back-seater for the first Pacific P-61 kill to later occur near Saipan.

In subsequent interceptions it became clear that Rabaul's bomber crews learned from their encounter with Bennett. When P-70s closed on Betty formations, the prey simply outran the attacker by entering a shallow dive. As a result, the P-70s changed tactics and patrolled outer perimeters under radar coverage while the unit's P-38s lurked within the inner attack zone aided by searchlights.

Back in New Guinea, Detachment A's first success took place about a month later on 15 May 1943 and was credited to Second Lieutenant Burrell Adams. Port Moresby had received ample warning of the attack when two outposts reported bogeys headed its way at 1910 that evening. Ground radar tracking commenced eight minutes later, however, the Japanese formation of six No. 702 *Kokutai* G4M1s overflew the town out to sea to attack from the reverse direction. These six led by Lieutenant Sumita Shinshichi had departed Vunakanau airfield at 1600, intent on bombing Port Moresby's airfields in three sections of two aircraft each. The Japanese operations log notes that Sumita's wingman returned fire with 127 x 20mm and 620 x 7.7mm rounds before Sumita was shot down. Sumita's pair made several runs over the target at 12,000 feet in an effort to ensure bombing accuracy. In doing so, he became more vulnerable when Adams closed on the pair after they had dropped their bombs over Durand 'drome. Adams' shooting was accurate, and the bomber crashed into jungle 35 miles northwest of Seven Mile 'drome at 2005.

On 7 May 1943, Detachment A's commander Captain Hornaday was lost returning from an instrument training flight south of Port Moresby. As he descended through low overcast observers at an Australian anti-aircraft battery at Pyramid Point observed his P-70 stall and explode into the sea just offshore. Hornaday was presumed to be a victim of an unsuccessful transition from instrument to visual flight.

On 22 November 1943 Detachment B also lost its commander, Captain John McCloskey. He was returning from a night training flight and crashed into the ocean at 0400 just short of Henderson Field on approach. Two crew survived but a few days later pilot Second Lieutenant Dean Vanderhoff went missing in a P-70 off Bougainville. The wreckage was later discovered on the island's west coast. On 15 December 1943 both detachments were recalled to John Rogers Field, Hawaii, to prepare for transition to the P-61 Black Widow. Both detachment's P-70s were amalgamated into the 418th and 419th NFS with nearly all personnel rotated back to the US.

A logo design with a skull insignia and a propeller background was approved in 1924 and was a legacy logo from the days of the 6th Pursuit Squadron. It was replaced by the Hoot Owl insignia carrying a machine gun when the unit converted to night fighters. Both logos are illustrated at the top of page 22.

P-70 serial 39-768 Dusty, the subject of Profile 2, taxies at Three-Mile 'drome in early March 1943 shortly after Detachment A arrived in New Guinea.

Pictured is First Lieutenant William McIvor who was a P-70 pilot with Detachment B at Guadalcanal. He later featured as the main actor in P-61 training films when he returned to the US to become a test pilot and instructor.

PACIFIC PROFILES

6th Night Fighter Squadron (Detachments A and B)

Profile 1   P-70 serial 39-766 Detachment A

The starboard fuselage illustrated a variation of the Hoot Owl squadron logo, with the owl wearing boxing gloves instead of carrying a machine gun. The aircraft was assigned to First Lieutenant Warren Thompson at Port Moresby.

Profile 2   P-70 serial 39-768 *Dusty* Detachment A

On 29 March 1943 *Dusty* was written off following a belly landing at RAAF Garbutt, Townsville, in north Queensland. The aircraft was on a ferry mission to bring back supplies to Port Moresby. The same stalking black leopard motif also appeared on serial 39-789 *Gloria* (see Profile 5).

Profile 3   P-70 serial 39-776 *Black Bug* Detachment B

This P-70 was among the first delivered to Detachment B when it joined the Thirteenth Air Force. It was photographed at Guadalcanal in April 1943.

Profile 4   P-70A-2 serial 42-54268 Detachment A

This new P-70 was assigned to Detachment A's commander Captain Warren Hornaday at Port Moresby. The motif illustrates a six-armed woman with a variety of weapons including a pistol and cleaver.

Profile 5   P-70 serial 39-789 *Gloria* Detachment A

Gloria was written off due to airframe damage when it was taxied into a ditch at Seven Mile 'drome, Port Moresby, on 7 September 1943. It was then towed to the airfield's boneyard where it was used for spares. Note it carries the same leopard motif as serial 39-768 *Dusty* (see Profile 2).

The wreck of P-70 serial 39-789 Gloria, as illustrated in Profile 5, after it was towed to Port Moresby's boneyard.

P-61A serial 42-5528 Jap Batty, as illustrated in Profile 6, during a stopover on the 1944 6th NFS ferry flight to Saipan. Note the radar antenna is just discernible though the opaque fibreglass radome.

P-61A 42-5554 The ~~Virgin~~ Widow, as illustrated in Profile 11, seen on Saipan.

CHAPTER 3
6th NFS (Reformed)

As explained in the previous chapter, Detachments A and B of the 6th NFS finished their respective tours of New Guinea and Guadalcanal in December 1943 and went to Hawaii. Some months later in April 1944 the 6th NFS, now reassigned to the Seventh Air Force, took delivery of its first eight P-61A Black Widows. These were shipped to Hickam Field where they were reassembled by the Hawaiian Air Depot and modified for Pacific operations. Once reassembled, they were flown to John Rogers Field where aircrews began familiarising themselves with the advanced fighter and its layout. Aerial gunner Staff Sergeant Leroy Miozzi became the unofficial unit artist and decorated seven of these P-61s with nose art shortly after delivery. These were 42-5522 *Dark Mistress*, 42-5523 *Japanese Sandman*, 42-5524 *Midnight Mickey*, 42-5525 *Midnight Belle*, 42-5526 *Nightie Mission*, 42-5527 *Moonhappy* and 42-5528 *Jap Batty*. One fighter, 42-5529, remained unnamed at that time.

After the squadron's pilots checked out in the P-61, the turrets were replaced with modified B-24 fuel tanks so they could ferry the P-61s to Saipan via Palmyra, Canton Island, Tarawa, Kwajalein, Eniwetok and finally to Saipan's Isley Field. The long ferry flight was led by unit commander Major Victor Mahr. The crews arrived in Saipan on 21 June 1944, from where they operated until 1 May 1945.

The 6th NFS logo remained the Hoot Owl insignia carrying a machine gun from the days of Detachments A and B but did not appear on any of the unit's P-61s.

P-61A serial 42-5527 Moonhappy, the subject of Profile 8, with its revised nose-art. Part of the original art can be seen to the right of the moon's tip.

PACIFIC PROFILES

6th Night Fighter Squadron

Profile 6 P-61A serial 42-5528 *Jap Batty*

This P-61A was received by the 6th NFS on 21 April 1944 at John Rogers Field, Hawaii. The original crew were pilot Second Lieutenant Jerome Hansen, radar operator Second Lieutenant William Wallace and gunner Staff Sergeant William Anderson. This crew was credited with a Betty on 8 July 1944. The fighter was written off on 3 November of that year when the starboard maingear collapsed on landing.

Profile 7 P-61A serial 42-5524 *Midnight Mickey*

Midnight Mickey was the first Black Widow to arrive on Saipan. Its first crew consisted of pilot Second Lieutenant Myrle McCumber, radar operator Flight Officer Daniel Hinz and gunner Private Peter Dutkanicz. The fighter survived the war and was condemned for salvage on 31 August 1945.

Profile 8 P-61A serial 42-5527 *Moonhappy*

Assigned to the crew of pilot First Lieutenant Dale "Happy" Haberman, radar operator Second Lieutenant Raymond Mooney and gunner Sergeant Patrick Farrelly, the aircraft's name combined the pilot's nickname and the radar operator's surname. This aircraft and crew were credited with four kills, including the first P-61 victory of the war on 30 June 1944. This was a No. 732 *Kokutai* G4M1 Betty flown by Warrant Officer Shozo Kohira. *Moonhappy* was scrapped at Kipapa Gulch airfield, Hawaii, on 31 August 1945. The aircraft had two nose-art versions, the first illustrated on the left.

Profile 9 P-61A 42-39573 *My Joan*

This P-61B was flown to Oakland for deployment on 1 January 1945 and was assigned into the 6th NFS on 9 March 1945. It went on to serve in postwar Japan.

Profile 10 P-61A 42-5526 *Nightie Vision*

Assigned to pilot Second Lieutenant Jerome Hansen and radar operator Second Lieutenant William Wallace, this pair were credited with downing a Betty on 8 July 1944. The fighter was ferried to Guam for salvage when the 6th NFS departed Saipan and was struck off charge there on 5 August 1945.

P-61A 42-5527 Moonhappy, as depicted in Profile 8, with pilot "Happy" Haberman and radar operator Raymond Mooney. This is the first version of the aircraft's nose artwork.

Profile 11   P-61A 42-5554 *The Virgin Widow*

On 4 June 1944, this Black Widow was assigned into the 6th NFS and the crew of pilot Second Lieutenant Robert Ferguson, radar operator Second Lieutenant Charles Ward and gunner Staff Sergeant Leroy Miozzi. As unofficial unit artist, Miozzi painted the artwork on the plane. In the early hours of 26 December 1944 this crew claimed a Betty over Kagman airfield on Saipan. The fighter was scrapped in 1946.

Profile 12   P-61A 42-5598 *Sleepy Time Gal II*

This P-61A was received by the 6th NFS on 22 June 1944 and was allocated to the crew of pilot Captain Ernest Thomas, radar operator Second Lieutenant John Acre and gunner Sergeant Leon Brill. This crew claimed a Betty on Christmas night 1944, and a second while detached to the 548th NFS on the night of 25 March 1945 with a gunner from that unit. The airframe was scrapped in 1946.

P-61A 42-5526 Nightie Vision, *as shown in Profile 10, showcasing one of artist Staff Sergeant Leroy Miozzi's more extravagant works.*

P-61A 42-39573 My Joan, *as illustrated in Profile 9, seen during postwar service with squadron number "10" on its nose.*

P-38G serial 42-12851, as shown in Profile 17, at Port Moresby in early 1944. (courtesy Chris Narzisi)

Major William "Bill" Sellers on the left with P-61A 42-39369 Ally R. This fighter is depicted in Profile 20 with the 418th NFS King Bee logo on its nose, but that had not yet appeared when this photo was taken.

CHAPTER 4
418th Night Fighter Squadron

The 418th NFS eventually became the top-scoring USAAF night fighter squadron in the Pacific. Activated on 1 April 1943, the unit's ground echelon docked at Brisbane on 29 October 1943 aboard the USS *General John Pope*. It then proceeded via Townsville to Dobodura, where it arrived on 22 November. At this time five P-70s from Detachment A of the 6th NFS were incorporated into the unit, and by February 1944 it had acquired five more P-70A-2s. Commanded by Major Carroll Smith, the 418th NFS answered directly to Fifth Fighter Command. To supplement the limited number of P-70s, a dozen P-38F and -G Lightnings were also acquired.

On 28 March 1944 the unit moved to Finschhafen, but two months later in advanced west to Hollandia and then in June it was in Wakde. The moves continued in quick succession to Owi and Morotai, before reaching the Philippines in November 1944. Major William "Bill" Sellers took over from Smith as commander on 15 January 1945.

The 418th NFS deployed its P-38s for night alerts and patrols from Dobodura, Port Moresby, Kiriwina and Finschhafen during its time in New Guinea, but opportunities for engagement were limited. The unit also conducted daytime missions during which it lost its first P-38 on 16 January 1944 when it crashed during a strafing run. On 2 January 1944 two P-38s patrolled over the amphibious landing at Saidor for two hours from dawn.

Tropical weather claimed more losses during extended patrols when using twin drop-tanks. For example, on 5 March 1944 Lieutenant Edward Craig went missing in the Admiralty Islands after running out of fuel before he could find an Allied base.

From 1 to 13 April 1944 the 418th NFS used its P-38s to run an extensive series of night patrols between Cape Croiselles and Kar Kar Island while temporarily based at Saidor. Whilst the occasional Japanese bomber flew overhead in the form of a Ki-48 Lily or Ki-49 Helen, these were transitory appearances *en route* to attacking Finschhafen. Saidor itself was rarely a night-time target in its own right. On the odd occassions that enemy aircraft were detected, there was rarely sufficient warning to scramble. With few worthwhile targets, and little hope of catching one even if it materialised, the P-70s were thus reduced to garrison duties during their Saidor deployment which included supply and leaflet dropping sorties. Overall, during its brief four months operating in New Guinea the 418th NFS lost four P-38s to accidental causes including one to "friendly" anti-aircraft fire.

With still no P-61s at hand, the unit next converted a dozen B-25Hs into night fighter configuration at Nadzab in March 1944. They removed the 75mm cannon and upper turret and converted them to strafer configuration with fourteen forward-firing machine guns. With few Japanese aircraft in New Guinea skies at this juncture of the air war, the Mitchells reverted to daytime strikes instead. Targets were mainly around Wewak but there was also the occasional night mission over Rabaul. In May 1944 the unit moved from Nadzab to Hollandia where shortly thereafter it commenced conversion to the P-61 in mid-August 1944 whilst still operating the B-25Hs.

On 16 September 1944 the air echelon flew their new P-61s to Owi where it collaborated with the 421st NFS in defending the island. In early October the P-61s moved to Morotai, along with three Ground Controlled Interception radar controllers, where the squadron established a system of night defence. Over Morotai on the night of 7 October 1944, the squadron claimed its first P-61 victory.

Ten replacement combined pilot/radar operator crews arrived in January and February 1945. These crews had acquired around 100 hours on the P-70 and 50 on the P-61 back in the US. Night patrols soon commenced from Mindoro during the Battle of Manila. These patrols were multi-purpose flights for hunting enemy aircraft, tracking enemy ground convoys and artillery spotting. Missions in March 1945 comprised mostly convoy and beachhead patrols, including over US landings at Lubang and the Palawan Islands. Commencing in April 1945, a technique for bombing by radar was developed. In June several missions to Formosa and along the China coast demonstrated the P-61's extended range using external drop fuel tanks.

The 418th NFS P-61s departed Mindoro on 25 July for Okinawa to participate in the air offensive against Japan. A series of intruder missions to Kyushu commenced on 28 July. On the night of 7 August, the squadron shot down a Betty bomber in the traffic pattern at Kumamoto airfield, representing its last victory. A few weeks after Japan's surrender, the unit's aircrews commenced returning to the US.

The first and unofficial 418th NFS logo was a tomcat with blood on its paws which was created during the P-70 era. Later some crews were decorating their P-70s with differing versions of a landing eagle instead. The squadron later adopted the official "King Bee" logo. This featured a black and golden orange bee, wearing a red crown and holding aloft a lighted lantern which represented radar. The left foreleg held a machine gun while the bee tip-toed across a cloud formation, peering ahead against the background of a crescent moon and two stars. The logo changed again during the P-61 era with a simple moon and stars which also featured a reverse slant blue stripe with white piping. The reverse slant goes against conventional heraldry, signifying the unit's outlier status within the USAAF as a night fighter unit. The early unofficial tomcat logo and the "King Bee" logo are illustrated at the top of page 33.

The Western Electric microwave radar installed in the fiberglass nose radome of the P-61 was top-secret when introduced.

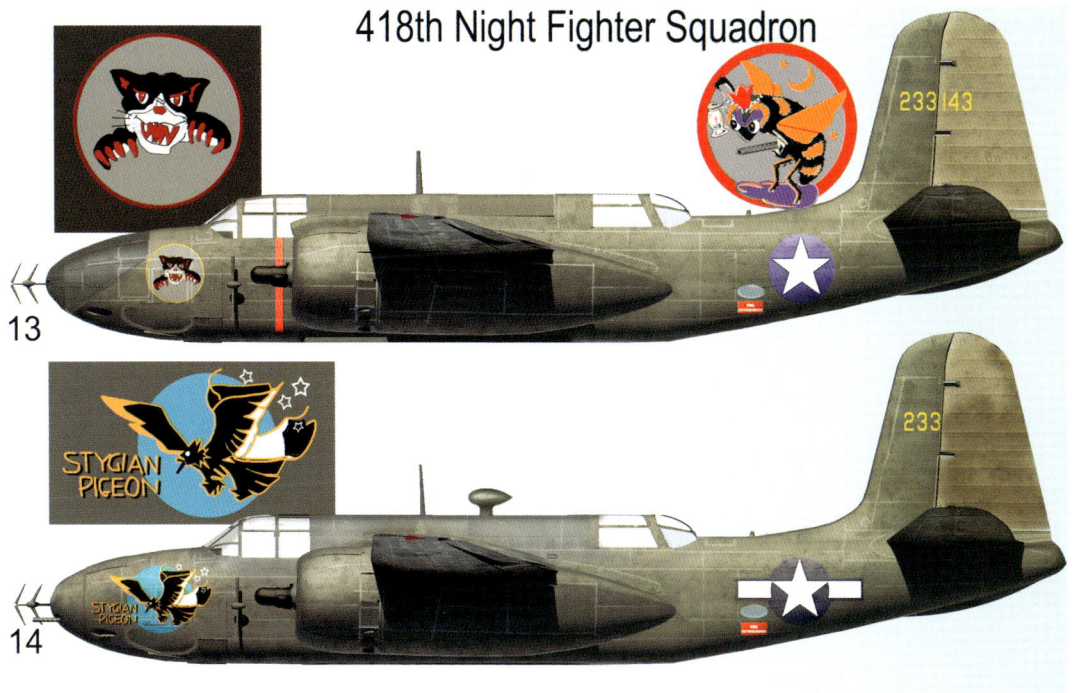

Profile 13   P-70A-1 serial 42-33143

On 2 July 1963 an Australian patrol officer surveyed the wreckage of this P-70A in the Astrolabe Ranges, about 1,000 ft above sea level. The nose section still sported the decoration of a grey circle outlined in red, "upon which a red animal or bird had been surmounted". USAAF records (then still classified) showed that the mystery wreck had disappeared after launching from Saidor airfield in the late afternoon of 2 April 1944 flown by Second Lieutenant Robert Forrestal, the third cousin to wartime Secretary of the Navy James Forrestal. The two-hour mission was to drop supplies to Australian soldiers and propaganda leaflets to local villagers, urging support for approaching Allied forces. Flight Officer Harold Holley's job was to jettison the leaflets through the bottom hatch of the rear gunner's compartment. Holley survived the crash and was cared for by villagers for a few days before he disappeared. It is likely he was captured and executed by Japanese forces.

Profile 14   P-70A-1 serial unknown, *Stygian Pigeon*

This fighter was named after the Stygian owl which habituates the west coast of the US and northern South America.

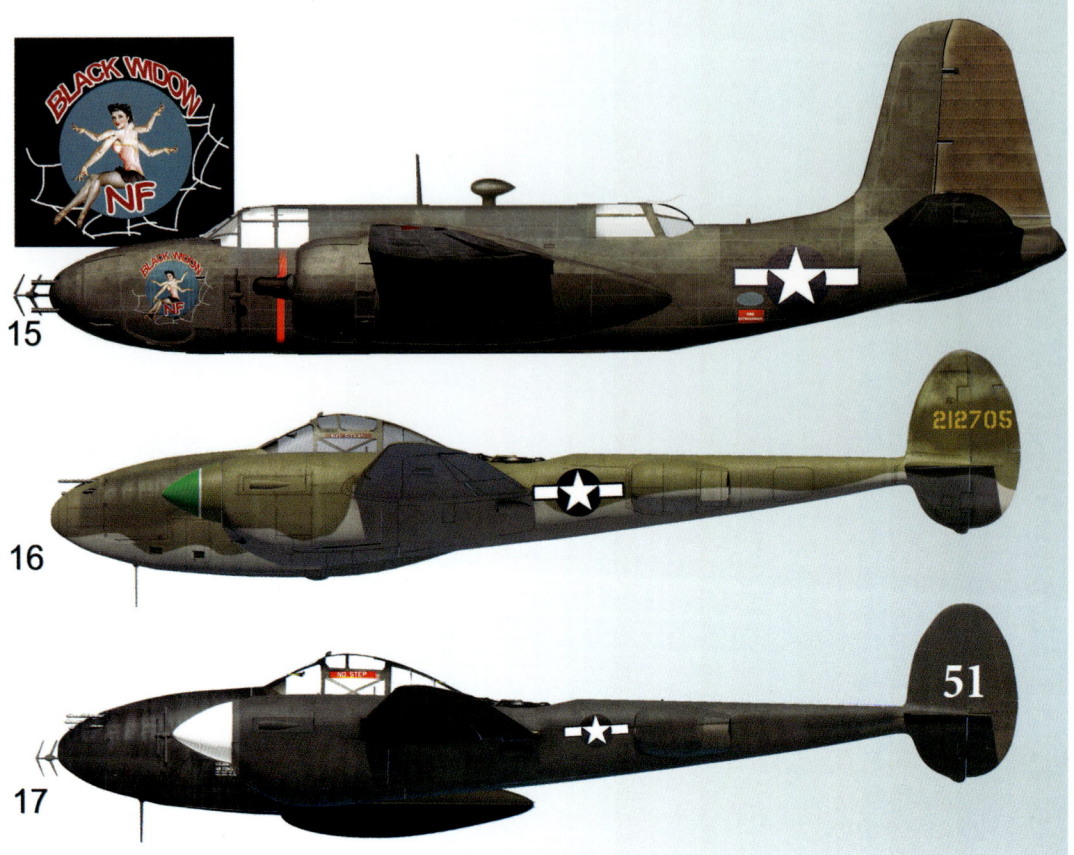

Profile 15   P-70A-1 or A-2, serial unknown, *Black Widow NF*

This P-70 appeared with the 418th NFS at Saidor in April 1944. "Black Widow" was a common colloquial phrase of the times, unrelated to the name later given to the P-61.

Profile 16   P-38G serial 42-12705

This Lightning was received from the 80th FS and remained in Olive Drab but had its name painted out. On 4 March 1944 Lieutenant Edward Craig went missing when caught in a storm during a patrol by a pair of P-38s between Manus Island and Finschhafen. The second Lightning flown by Second Lieutenant Robert Forrestal (see Profile 13 above) made an emergency landing at Cape Gloucester.

Profile 17   P-38G serial 42-12851, squadron number 51

This converted night fighter was received from the 6th NFS around December 1943 when that squadron returned to the US. It was later destroyed in an accident on 9 September 1944. The squadron number represented the last two digits of the serial number.

Profile 18   B-25H-1 serial 43-4422

On 27 July 1944, this Mitchell and three others departed Hollandia on a barge hunt to the east of the Japanese airfield complex at Babo flown by Lieutenant Ira Barnett. When Barnett was hit by anti-aircraft fire in the left engine, he experienced control problems and force-landed in kunai grass about 30 miles inland. A fellow B-25H located the downed bomber the next day and a rescue team was delivered to a nearby river-mouth by PBY. The team located the crew on 16 August, nearly three weeks after they had put down, and retrieved them back to the coast from where they were evacuated to Biak.

Profile 19   B-25H-1 serial 43-4158 *Blues in the Night*

Pilot First Lieutenant William "Bill" Ross often flew this Mitchell with radar operator Second Lieutenant Raymond Duethman. Ross named the bomber after the popular song of the times which had featured in the 1941 film released under the same name.

PACIFIC PROFILES

Profile 20   P-61A serial 42-39369 *Ally R*

This P-61A was assigned to 418th NFS commander Major William "Bill" Sellers with radar operator Lieutenant Hap Holiday. The King Bee logo against a white cloud was applied to the nose of this fighter towards the end of the war and can also be seen on Profile 23 below.

Profile 21   P-61B serial 42-39586 *Black Panther*

The crew of First Lieutenant Stan Logan and radar operator Second Lieutenant George Kamajian named this fighter which commenced service too late in the war to achieve any kills.

Profile 22   P-61B serial 42-39595 *Time's A Wastin'*

This Black Widow's original crew included squadron commander Major Caroll Smith and radar operator Second Lieutenant Philip Porter. Smith became the top-scoring P-61 pilot in the Pacific. His crew's claims in 1944 included a Dinah on 7 October, two Irvings on 29 December and a Rufe and Frank on 30 December.

Profile 23   P-61B Serial 42-39578 *Virginia L*

This Black Widow was received by the 418th NFS on 29 February 1945 and assigned to pilot First Lieutenant Al Nielsen and radar operator Second Lieutenant Homer Piett. It was transferred postwar to the 4th All Weather Fighter Squadron at Naha Air Base, Japan, on 20 February 1947 where it was converted to F-61B status before it was decommissioned on 28 June 1950.

Profile 24   P-61B 42-39661 *Nosey Nighthawk*

This 418th NFS P-61B was assigned to First Lieutenant Max Barker and radar operator Second Lieutenant Richard Wood.

P-61B 42-39595 *Time's A Wastin'* at Morotai. This fighter is illustrated in Profile 22.

One of the early night fighters employed by the 419th NFS was P-70A-2 serial 42-54273. The subject of Profile 25, it is seen at Guadalcanal in late 1943.

A contingent of 419th NFS ground personnel pose with P-61A serial 42-5597 Little Feller, as shown in Profile 30.

CHAPTER 5
419th Night Fighter Squadron

On 15 November 1943, the newly formed 419th NFS arrived on Guadalcanal equipped with ground control radar but no aircraft. Six of the unit's fifteen pilots had previously served with the Royal Canadian Air Force in night fighter operations, and the squadron had been promised the new P-61 Black Widow upon arrival. However, it was assigned to the parent 18th Fighter Group and instead allocated hand-me-down P-70s from the 6th NFS Detachment B then in the process of returning to the US. With these it managed only three night patrols, six scrambles, four intruder missions and four daylight sorties by the year's end.

Demoralised by having to operate combat-weary P-70s, matters got worse when during its first few months it claimed no victories at a cost of five aircraft and four pilots. The 419th NFS also acquired Lightnings which they operated longer than envisaged, losing several to accidents while pilots became acquainted to their high-performance characteristics. A total of 368 night missions was eventually logged on the type about which the squadron historian recorded:

> … not once did our P-38s come close to an enemy intruder. Our gunners and radar observers were unemployed - "peeved" is a common word used around here.

When the limitations of the P-38 as a night interceptor became apparent, they were used alongside the P-70s for daytime harassment missions. These ranged as far as Bougainville and Rabaul, staging through Torokina. In April 1944 the 419th NFS also acquired several RA-24Bs, the USAAF reconnaissance version of the SBD dive-bomber. These were employed for instrument training, a useful diversion for the unit's bored pilots whilst waiting for the long-promised P-61 Black Widows.

The first P-61 arrived in the Solomons on 3 May 1944 when several were unloaded at Kukum wharf, located just off the end of Guadalcanal's bomber strip. P-61A serial 42-5506 was the first of these to fly, a test flight made from Guadalcanal on 12 May 1944 by squadron commander Major Emerson Barker.

The 419th NFS later painted one of their Lightnings yellow and black, and used it for target towing so that pilots of the newly assembled Black Widows could refine their gunnery skills (see Profile 86). On 26 June 1944 the squadron's P-61s were ferried to the massive airfield complex of Nadzab, in New Guinea's Markham Valley. Here the crews trained for an intense month, whilst sister squadrons the 418th and 421st NFS waited impatiently to receive their P-61s. The 419th NFS moved to the southern Philippines in 1945 where aerial targets were increasingly rare. The unit flew long-range attack missions and ended the war stationed on Palawan.

The 419th NFS allocated its first batch of P-70s and P-38s squadron numbers 301 through to 309, stencilled in white on the nose, with USAAF serials retained on the fins. Initially the Lightnings were left in the standard Olive Drab scheme, however, later the undersurface of some were painted matt

PACIFIC PROFILES

black (see Profile 28). The squadron numbers were recycled to different aircraft when airframes were lost. Numbers 310 to 321 were subsequently allocated to the first batch of P-61s at Guadalcanal. The squadron number was moved to the tail later in the war. The 419th NFS logo featured a critter resting on a cloud holding an illuminated lantern, as illustrated below.

Profile 25   P-70A-2 serial 42-54273

Several P-70A-1s which served with the 419th NFS such as this example had previously served as training aircraft in the US. These retained the last two numerals of the serial number painted in white on the tail.

Profile 26   P-70A-1 serial 42-33152 squadron number 308

This P-70A-1 was equipped with a SCR 540 radar and fitted with four antennas on each side of the fuselage. These were mounted in the horizontal azimuth, which produced a signal strength differential indicating to the operator the degree to which the target was offset. This aircraft was repaired following a landing accident at Guadalcanal on 8 January 1944.

Profile 27   P-38J serial 42-67168 squadron number 301

Illustrated as it appeared at Piva 'drome, Bougainville, in January 1944, this Lightning was later transferred to Combat Replacement Training Center at Nadzab. While serving with that unit it was lost in an accident on 12 May 1945.

Profile 28   P-38J serial 42-67788 squadron number 305

Illustrated as it appeared at Piva 'drome, Bougainville, in April 1944, the underside of this 419th NFS Lightning was painted matt black.

Ground personnel pose with P-61A serial 42-5515 Sleepy Time Gal, as depicted in Profile 32.

PACIFIC PROFILES

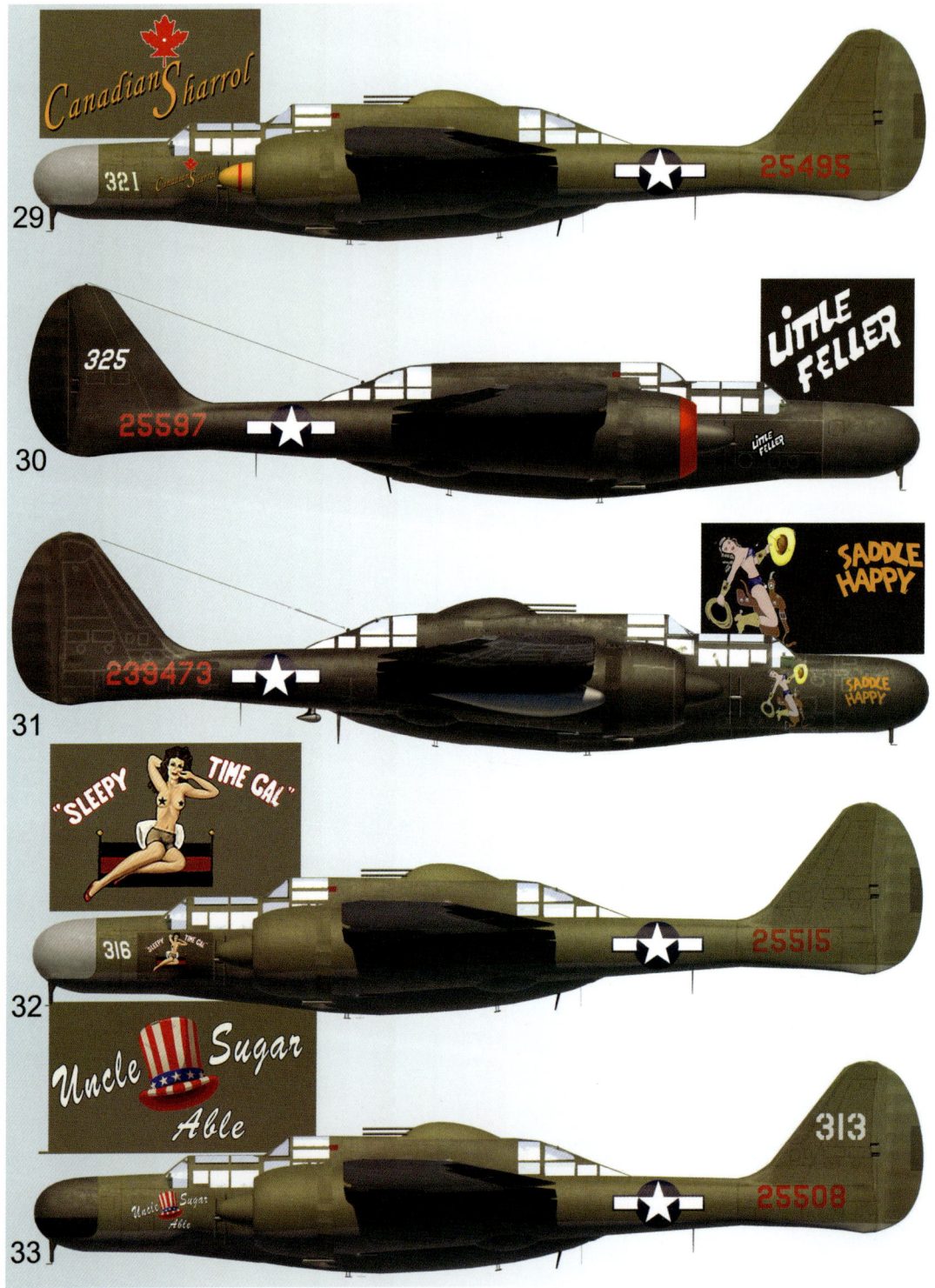

Profile 29   P-61A serial 42-5495, squadron number 321, *Canadian Sharrol*

This P-61A was a 419th NFS original assigned to Captain John Laffey on 6 May 1944 and given squadron number 321. Note the spinners were painted yellow with a red pin stripe. From Rockford, Illinois, Laffey named the aircraft Sharrol, a girl's name more common around Quebec. Laffey went missing on 6 March 1945 in another aircraft. This fighter continued service with other crews and was credited with a Kawasaki Ki-45 Nick on 13 June 1945. Soon afterwards the aircraft was ditched on 10 July 1945 after both engines quit. All three crewmen were rescued.

Profile 30   P-61A serial 42-5597, squadron number 325, *Little Feller*

Assigned into the 419th NFS on 13 July 1944, this P-61A suffered an accident on 1 October 1944 when a routine electrical check accidentally shorted out the cannon circuit breakers, causing one to fire. The round struck the nosewheel brace and doors, where it exploded and started a fire killing armourer Staff Sergeant Fred Straub. The aircraft was repaired and put back into service.

Profile 31   P-61B serial 42-39473 *Saddle Happy*

This P-61B was assigned to the 419th NFS on 5 January 1945 and the crew of pilot First Lieutenant Robert Frenzel, radar operator Second Lieutenant George Black and gunner Staff Sergeant Cecil Fuller who agreed to name it *Saddle Happy*. In their first-assigned P-61, Frenzel flew the longest mission ever flown by a P-61 on 17 September 1944, when he flew a 1,600-mile night intruder mission of just over nine hours duration to Borneo. *Saddle Happy* was later transferred to the 550th NFS where it was named *Night Vision* (see Profile 67)

Profile 32   P-61A serial 42-5515, squadron number 316, *Sleepy Time Gal*

The 419th NFS took delivery of this Black Widow on 18 May 1944. *Sleepy Time Gal* was named by the crew of pilot Captain Howard Daniel, radar operator Second Lieutenant Harold Ozmun and gunner Sergeant William Yahn. On 18 October 1944 it departed Sansapor for a dawn intruder mission around Waigeo Island. Scheduled to return at 0630, the aircraft went missing and its wreckage was found in jungle near New Guinea's Vogelkop Peninsula by an RAAF team postwar.

Profile 33   P-61A serial 42-5508, squadron number 313, *Uncle Sugar Able*

This P-61A was among the original batch unloaded at Guadalcanal in May 1944. Its first assigned crew was First Lieutenant Robert Oates, radio operator Second Lieutenant Paul Ferguson and gunner Sergeant Adolfo Falcon. The starboard propeller threw a blade through the fuselage during a mission on 8 December 1944, however, the aircraft was returned to service. It was stored at Clark Field after the war and scrapped on 11 October 1946. The name was radio code for the letters "USA".

Profile 34   P-61B serial 42-39450, squadron number 329

Assigned to the 419th NFS on 9 December 1944, this Black Widow, known informally as *Lady Luck*, was operated by pilot First Lieutenant Phillip Hans, radar operator Second Lieutenant "Doc" Holloway and gunner Sergeant Donald Clancy. Following its transfer to the 421st NFS the aircraft was lost in a take-off accident on 4 July 1945 in the Philippines. Piloted by First Lieutenant Loren Kane, the Black Widow failed to climb during a dawn departure. It hit the runway hard then skidded into the ocean where it caught fire. Kane lost his life, however, the other two crewmen escaped.

Profile 35   P-61A serial 42-5563, squadron number 324, *Old Salty Dog II*

Assigned into the 419th NFS in July 1944 at Guadalcanal, squadron operations officer Captain Howard Daniel subsequently ferried the fighter to Middelburg where it was assigned to pilot First Lieutenant Robert Winslow, radar operator Second Lieutenant Tony Grotzinger and gunner Sergeant MW Ridenour. Eventually credited with two Dinahs, *Old Salty Dog II* was condemned for salvage on 27 February 1945 at Clark Field. Note the two pin stripes on the spinner and the decorated nosewheel.

A ground crew member hands a 419th NFS Black Widow pilot his parachute whilst others turn the propeller to prevent hydraulic lock just prior to start-up. The fighter is P-61A serial 42-5563 Old Salty Dog II and is illustrated in Profile 35.

A worn looking P-61B serial 42-39450, the subject of Profile 34. Note the squadron number 329 is painted on the undercarriage door.

P-61B serial 42-39546 *Which Way'd He Go George?* seen with white-painted drop tanks. This Black Widow features in Profile 45.

P-61B serial 42-39439 *Nocturnal Nuisance*, as shown in Profile 47. This photo showcases the landing light added to the front oleo, a feature unique to all "B" models.

CHAPTER 6
421st Night Fighter Squadron

The 421st NFS arrived from the US at Townsville on Christmas Day 1943 before it quickly continued to Milne Bay, New Guinea. Commanded by Major Walter Pharr with operations officer Captain William Bradley, the unit arrived in New Guinea without aircraft. Pharr had been promised P-61s upon arrival, however, the new type would not materialise until mid-1944. By the end of January and still without aircraft, the unit was ordered to move to Nadzab.

The first aircraft received by the 421st NFS was a pair of combat-weary P-38Fs delivered to Port Moresby's Kila 'drome on 31 January 1944, followed by a solitary P-70 transferred from the 418th NFS a fortnight later. Four more ex-418th NFS P-70s followed in the next few weeks before one new P-70A-2 was ferried up from Brisbane. All required extensive maintenance before they were ready for combat at Saidor by the end of March.

The two P-38Fs were flight-tested from Kila 'drome then on 16 February two more P-38Gs were delivered, and a fifth three days later. On 21 February fourteen pilots, most of the aircrew cadre, took a C-47 to Townsville where they collected five new P-38s to ferry back to Nadzab. However, disappointingly these were commandeered by other combat units upon arrival. On 29 February the 421st NFS flew its first combat mission when three P-38Fs flew a patrol over the US landing in the Admiralties. Capricious weather saw one abort followed by the unit's first loss. This occurred when Lieutenant Paul Zimmer's gear would not retract fully, then his port engine governor failed. Finally, his starboard engine lost power which saw him parachute about 50 miles up the coast from Finschhafen, but he was safely back at base that evening.

Another curious loss occurred on 24 March; for the past several weeks ace pilot Lieutenant Richard Bong had accompanied the unit's fighters on reconnaissance missions and for this morning's mission Lieutenant Tom Malone borrowed Bong's decorated plane *Marge*. The mission was a two-aircraft patrol to Wewak, and half an hour into the flight at 30,000 feet Malone experienced engine trouble. Unable to feather the propeller, he baled out high above overcast and unclear of his position. The identity of the wreck was confirmed at the crash site by Justin Taylan of the Pacific Wrecks website in 2024.

Meanwhile five more P-70A-2s had been assigned into the 421st NFS, after which on 5 April Captain Bradley led six P-70s comprising A Flight to Saidor. From here the detachment commenced a series of low-level attacks against targets along the northern New Guinea coast. Back at Nadzab on 8 April 1944 Lieutenant Alexander Kuzmick became the unit's first fatality resulting from a landing accident in a P-38F.

On 21 April Major Pharr flew to Brisbane to inspect the first four P-61As assigned to the 421st NFS. A few days later, with the go-ahead from Pharr, fifteen pilots boarded a C-47 and made their way to Brisbane where the four fighters were in the process of being assembled. Northrop Corporation's chief test pilot, John Myers, was there to assist with the training program for conversion to the new type. In the midst of this Pharr suffered medical problems and he was returned to the US for rest. The operations officer, Captain Bradley, assumed acting command.

A further three P-70s joined the 421st NFS at Nadzab in late April. These enabled the formation of C Flight tasked with weather reconnaissance duty, sometimes flying two missions per day. On 28 May four P-70s were transferred to Wakde to intercept Japanese bombers which had been conducting night harassment raids. The squadron finally took delivery of its first three P-61As at Nadzab on 1 June, with another two arriving a week later.

In October 1944 the 421st NFS moved to the Philippines, setting up at San Marcelino in February 1945. The squadron saw out the war from there with both enemy interceptions and bombing missions.

In June 1943 the wife of Chief Engineering Officer Lieutenant John Olley won a competition in the US to design the best squadron insignia. The prize for the successful "Mad Rabiteers" logo was dinner for two at the Orange Court Hotel in Orlando, Florida. Mrs Olley's original drawing depicts Bugs Bunny riding a P-70, as depicted on the top of page 49. The 1945 version replaced the P-70 with a P-61 Black Widow.

P-61A serial 42-5502, as illustrated in Profile 40, carried nose art on both sides of the airframe. The starboard art of Nocturnal Nemesis is seen here.

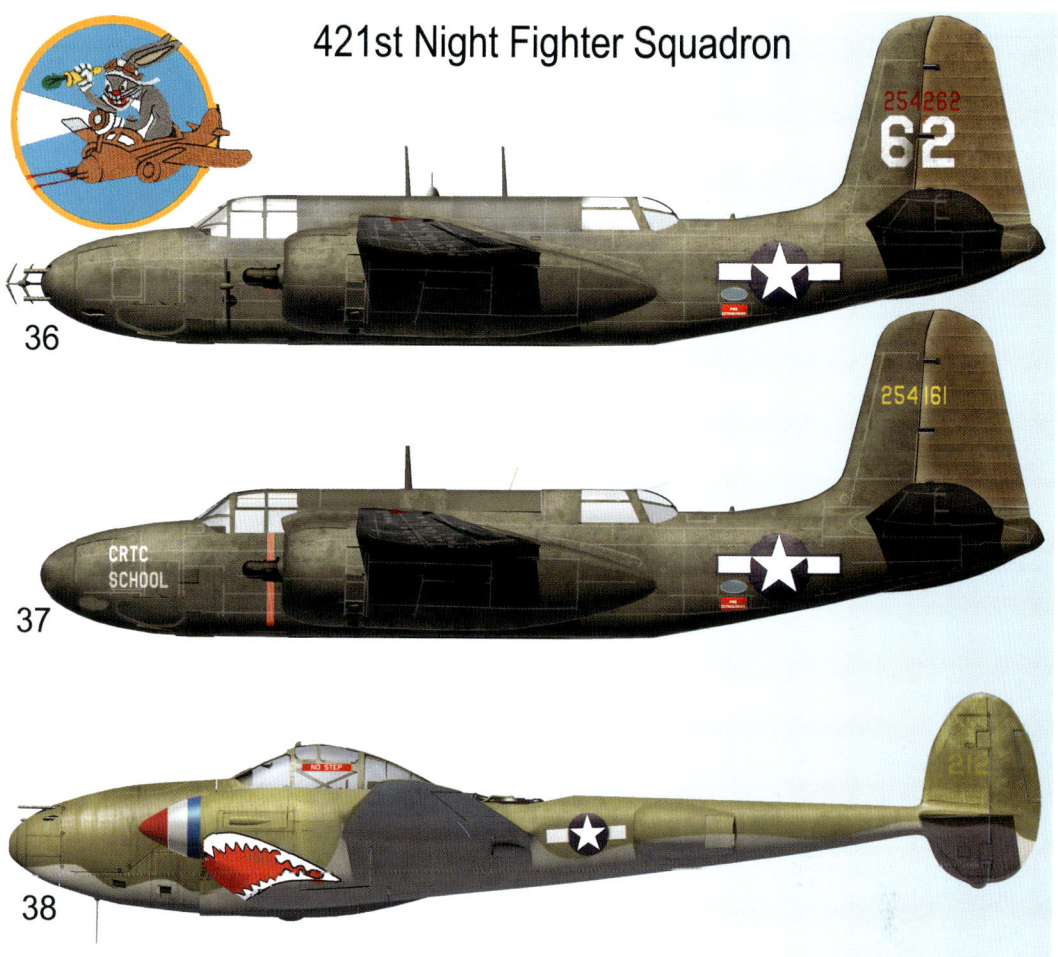

Profile 36 P-70A-2 serial 42-54262, squadron number 62

This P-70A was transferred from the 418th NFS at Nadzab. The last two numbers of the serial were enlarged in white on the fin to assist with aerial identification.

Profile 37 P-70A-2 serial 42-54161

This P-70A was transferred to the 360th Service Group (Combat Replacement Training Center) at Nadzab in late June 1944 when the 421st NFS commenced conversion to P-61s. On 22 October 1944 it went missing in the Mount Albert Edward area flown by Captain Donald Sutliff with two others aboard. It was last sighted by a P-47D mid-afternoon cruising at 14,000 feet on a ferry flight from Port Moresby to Nadzab. It is illustrated as it appeared with the CRTC, as stencilled on the nose.

Profile 38 P-38F serial unknown

This Lightning was transferred to the 421st NFS at Nadzab from the 9th FS. The spinners were painted in the patriotic colours of the US flag, and a unique shark's teeth design was applied around the engine intakes by the ground crew.

PACIFIC PROFILES

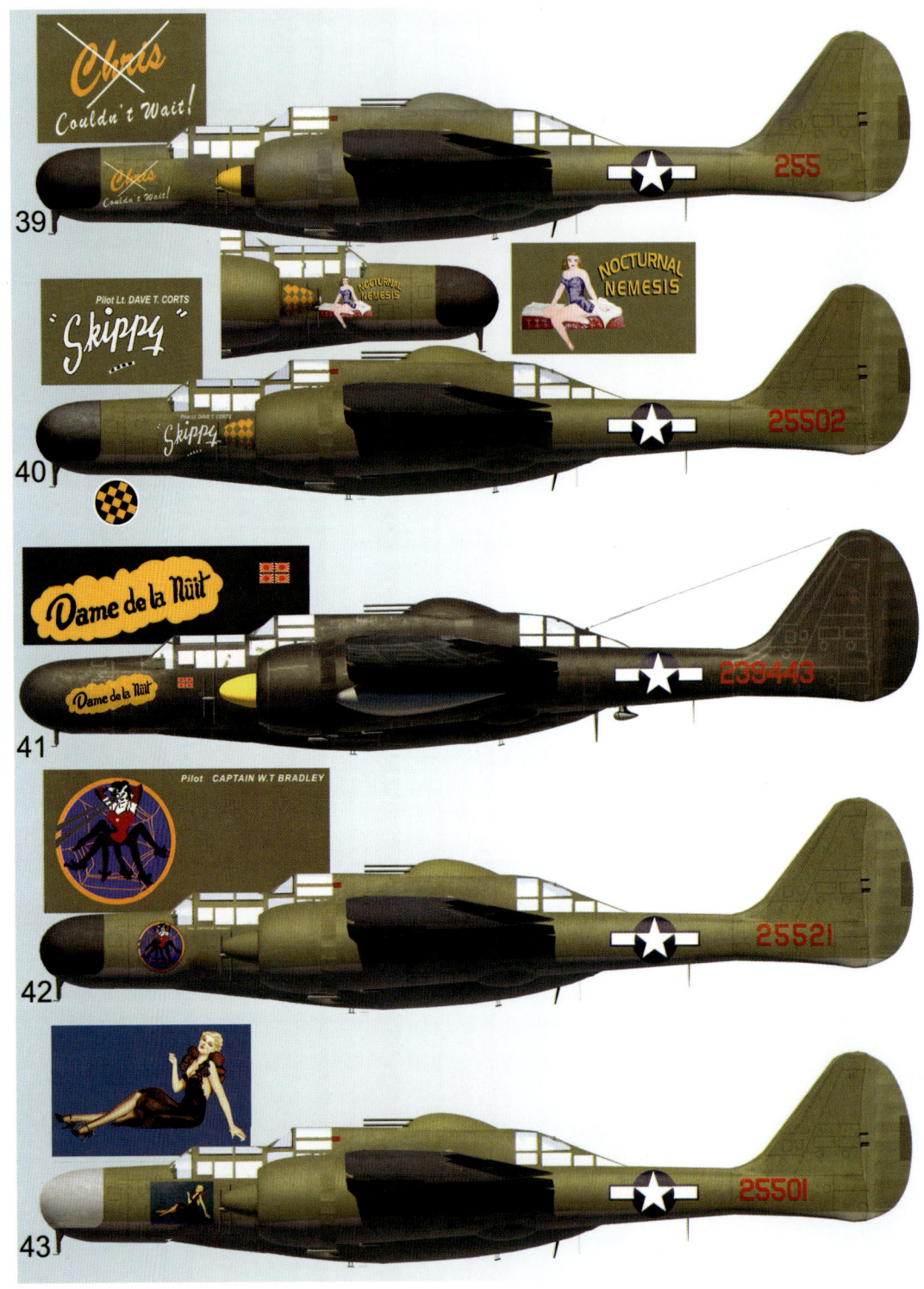

Profile 39   P-61A serial unknown *Chris Couldn't Wait!*

This Black Widow was assigned to First Lieutenant Paul Zimmer, a 421st NFS original who participated in the unit's first combat mission. On 22 April 1945 the fighter burned to the ground when it caught fire during a maintenance accident. Zimmer named the aircraft after a former girlfriend.

Profile 40   P-61A serial 42-5502 *Skippy / Nocturnal Nemesis*

This P-61A was assembled at Eagle Farm, Brisbane, in April 1944. It was one of the first four P-61As assigned to the 421st NFS and was delivered to Nadzab on 1 June 1944. There it was assigned to pilot First Lieutenant David Corts who named it after a girl he knew in Florida. The other crew were radar operator Second Lieutenant Alexander Berg and gunner Staff Sergeant Millard Braxton. On the night of 9 September 1944 Corts was credited with downing a Betty. Whilst parked on Tacloban's flightline, *Skippy* was damaged beyond repair during a bombing raid on 15 November 1944.

Profile 41   P-61B serial 42-39443 *Dame de la Nuit*

This "B" model left the factory without a top turret, however, one was fitted in the field after it was assigned into the 421st NFS on 18 December 1944. In late January 1945 it was assigned to First Lieutenant Owen Wolf, radar operator Second Lieutenant John Owen, radar operator Second Lieutenant Byron Allain and gunner Sergeant Donald Trabing. Wolf decorated the nose with four victory markings and the nose art which is French for "lady of the night". On 8 April 1945 this fighter was written off in a take-off accident at Clark Field. Note that this P-61B has elsewhere been misidentified as P-61A 42-5516.

Profile 42   P-61A serial 42-5521 (*Spider Queen*)

After being ferried to Brisbane from Hawaii, this Black Widow was assigned into the 421st NFS in May 1944 and the crew of pilot Captain Bill Bradley, radar operator Second Lieutenant Alton Woodring and crew chief Staff Sergeant Joseph Bradford. Bradley chose the nose-art and although it carried no name it was known along the flightline as either "the Queen" or "Spider Queen". It was scrapped at Clark Field in 1946.

Profile 43   P-61A serial 42-5501 (*sitting lady*)

This unnamed P-61A was assigned into the 421st NFS at Nadzab in early June 1944. On 28 November 1944 whilst based at Tacloban its crew downed a Zero which opened fire from behind. On 3 May 1945 when landing at San Marcelino Field, Luzon, one of the maingears collapsed causing the aircraft to belly land, and it was condemned due to structural damage. The pilot, First Lieutenant Hoke Smith, had the aircraft decorated with a calendar girl, however, no name was applied.

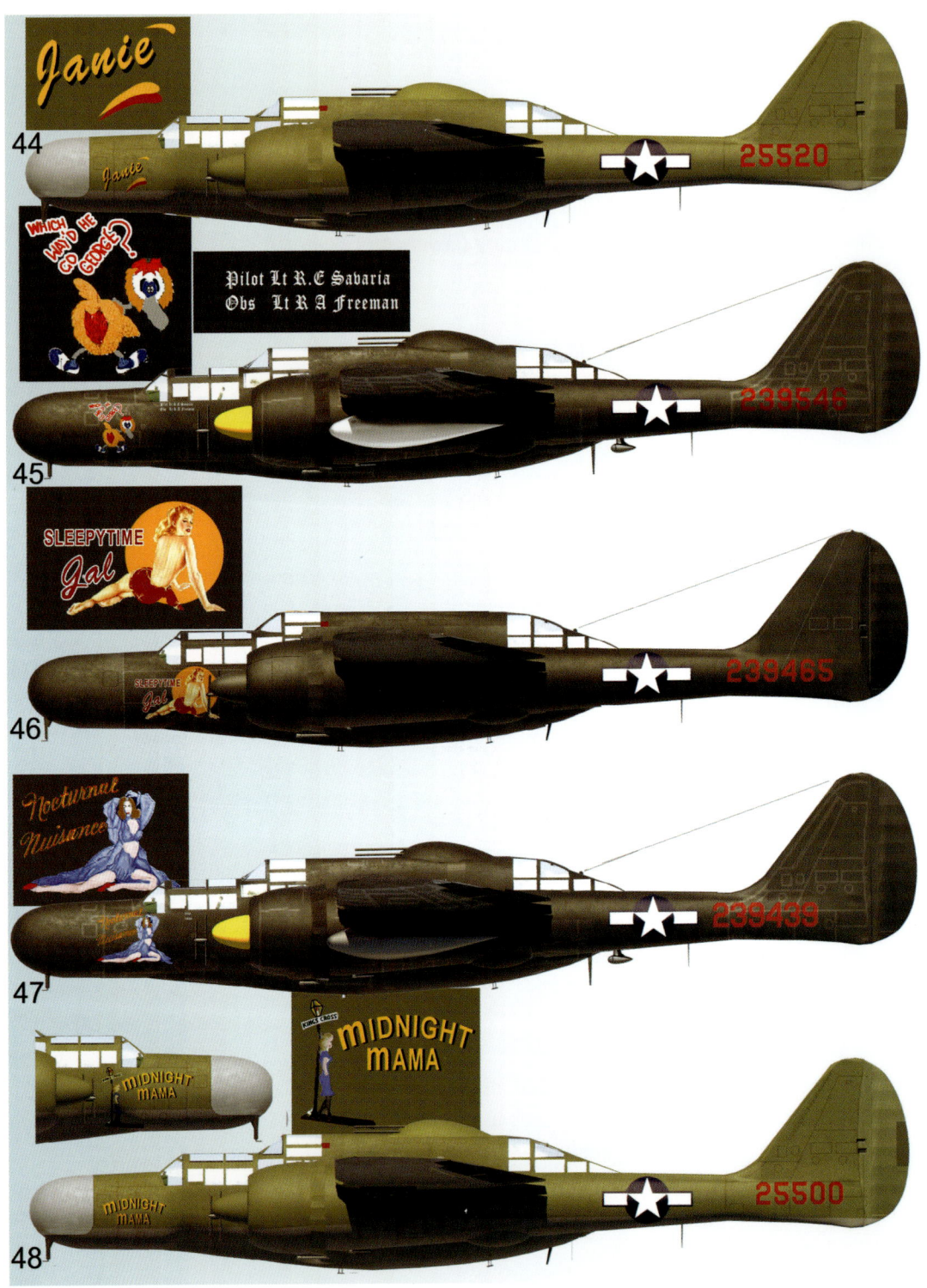

Profile 44   P-61A serial 42-5520 *Janie*

This Black Widow was credited with a Ki-84 Frank on 16 November 1944 over the Surigao Straits whilst covering PT boat operations. On 21 February 1945 a loss of hydraulic pressure caused a hard landing which broke both tail booms and the airframe was written off.

Profile 45   P-61B serial 42-39546 *Which Way'd He Go George?*

This "B" model had perhaps the most creative nose art in the 421st NFS which emulated the wartime artwork of cartoonist Tex Avery. The name and art were the invention of assigned pilot First Lieutenant Robert Savaria who would often pose the question to his radar operator Second Lieutenant George Freeman. Savaria escorted the Japanese Peace delegation back to Japan from Ie Shima after pre-surrender negotiations had been conducted. On 23 August 1963 the now civilian pilot Savaria was conducting fire-bombing over California in the world's last flyable P-61 when the starboard wingtip hit trees, cartwheeling the firebomber and killing Savaria.

Profile 46   P-61B serial 42-39465 *Sleepytime Gal*

Carrying the most ubiquitous name for USAAF aircraft in WWII, this one was named by assigned pilot First Lieutenant Fred Kohl with radar operator Second Lieutenant Robert Kunzman. The aircraft was assigned into the 421st NFS on 9 January 1945 but was destroyed on 10 July 1945 by flying debris from a B-24 crash landing at Clark Field.

Profile 47   P-61B serial 42-39439 *Nocturnal Nuisance*

This Black Widow was assigned to the commander of A Flight, First Lieutenant Tom Malone. It was being flown by a different pilot on 3 May 1945 who overshot the runway on his third landing attempt. The badly damaged fighter was stripped for parts and condemned for salvage.

Profile 48   P-61A serial 42-5500 *Midnight Mamma*

Shortly after being assembled at Townsville in May 1944 this P-61A was decorated by squadron artist Sergeant Rubin Kaplan for assigned pilot First Lieutenant Al Lockard. Kaplan went on to paint most of the 421st NFS's Black Widows including *Skippy / Nocturnal Nemesis*, *Bright Eyes*, *Nocturnal Nuisance* and *Spider Queen*. Kaplan was killed on 4 November 1944 during a bombing attack whilst he was standing in line for breakfast. *Midnight Mamma* was destroyed on 16 January 1945 at Tacloban, Leyte, when a B-24 bomber blew a main tyre on landing and struck the P-61.

Colonel Robert Johnson (far left) poses with P-61A serial 42-39390 Hard to Get, as illustrated in Profile 49. Johnson was not the regular pilot but claimed a victory on 28 February 1945 which was recognised by the freshly painted Japanese naval flag under the cockpit. The other three men are the regular crew of (second from left to right) pilot Captain Ken Scheiber, radar operator Lieutenant Bonnie Rucks and crew chief Technical Sergeant John Crough.

The same crew line-up as above (in reverse) this time in front of the starboard side of P-61A serial 42-39390 Hard to Get showcasing the nose-art.

CHAPTER 7
547th Night Fighter Squadron

The 547th NFS was the last P-61 squadron to join the Fifth Air Force, following on the heels of the 6th (Detachment A), 418th and 421st NFS. It was established on 1 March 1944 at Hammer Field, California, and was the first night fighter squadron to be established in that state (all the previous night fighter squadrons had been formed in Florida). Its P-61s operated throughout various airfields in the San Joaquin Valley before deploying to Owi via Dobodura in New Guinea in late August 1944.

The 547th NFS operated P-38s for its first month of operations until the arrival of the first P-61 in September 1944. The squadron operated small several detachments of two or three P-61s to enable wider coverage of interceptions. In January 1945 the unit moved to Lingayen on Luzon where the P-61s were field-modified to carry wing ordnance, including bombs, napalm tanks and rockets. This enabled the 547th NFS P-61s to conduct attacks against enemy airfields. The squadron's final move was to Ie Shima in preparation for the anticipated US invasion of Japan. Following the Japanese surrender it moved to Atsugi near Tokyo in October 1945 as part of the occupation forces. It remained there until February 1946 when it was deactivated.

The unofficial logo of the 547th NFS used late in the war featured a spider holding a variety of weapons including a pistol, musket, bomb, axe, mace and a knife. The logo is illustrated at the top of page 56.

A crewmen in an open front cockpit hatch of a P-61.

547th Night Fighter Squadron

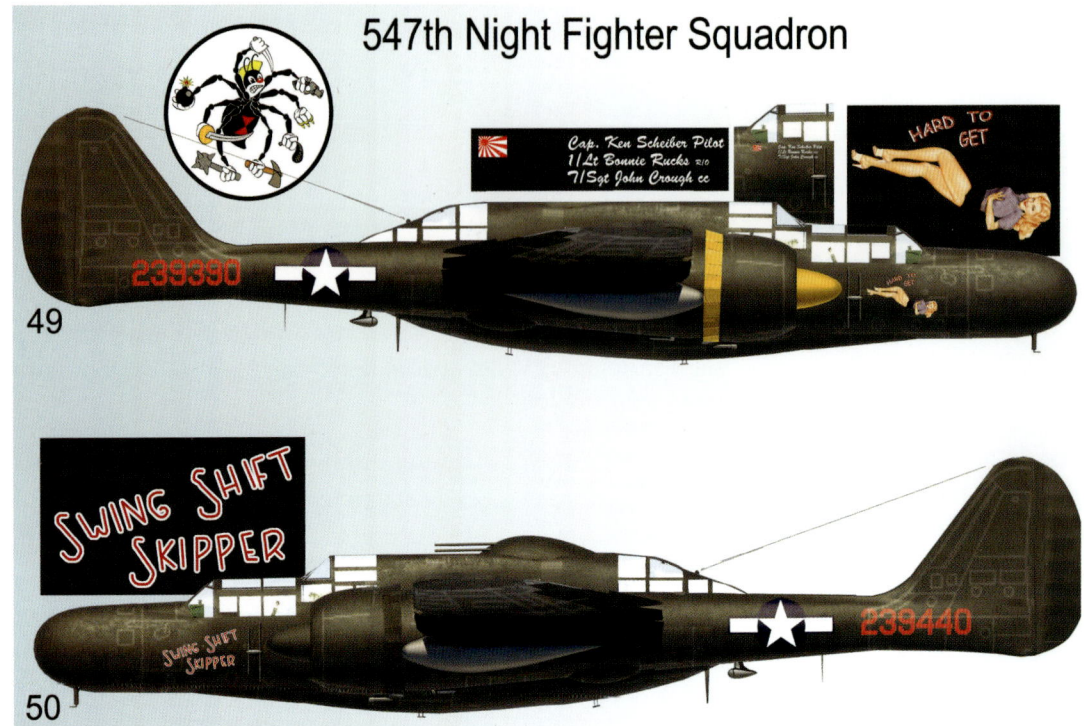

Profile 49   P-61A serial 42-39390 *Hard to Get*

This late "A" model without a turret featured a Japanese flag on the port side symbolising a victory awarded on 28 February 1945 when Colonel Robert Johnson, an observer from wing headquarters, claimed a Tess transport when it landed at night at Cagayan. The fighter's original crew were pilot Captain Ken Scheiber, radar operator Second Lieutenant Bonnie Rucks and crew chief Technical Sergeant John Crough. Rucks later transferred to the crew of Profile 50 (below). The fighter's cowls and spinners were painted yellow.

Profile 50   P-61B serial 42-39440 *Swing Shift Skipper*

Assigned into the 547th NFS on 6 November 1944, the fighter's first assigned crew was First Lieutenant Arthur Bourque, radar operator Second Lieutenant Bonnie Rucks and crew chief Sergeant Walter Nowak. Bourque had trained with the Royal Canadian Air Force in 1940 and subsequently flew with the RAF as a night fighter pilot in England. This crew was awarded two Betty bombers on the night of 19 February 1945. The aircraft was condemned on 28 September 1945 and was later scrapped at Clark Field.

The calligraphy on P-61B serial 42-39440 Swing Shift Skipper (Profile 50) typified the style on several other 547th NFS Black Widows being thin red lettering with wider white piping.

A ground crewman makes final adjustments on the dorsal turret gun pack of a Black Widow.

The 548th NFS logo and artwork on P-61B serial 42-39408 *Lady in the Dark*, as featured in Profile 56. The fighter is in disrepair and waiting to be scrapped.

The 548th NFS flight line on 10 March 1945. Several Black Widows carry the squadron logo under their cockpit.

CHAPTER 8
548th Night Fighter Squadron

Established on 23 March 1944, the 548th NFS was one of the last dedicated USAAF night fighter squadrons to deploy to the Central Pacific. It first trained at airfields located in the San Joaquin Valley before deploying to Hawaii in September 1944 where it was assigned to the Seventh Air Force. After contributing to Hawaii's defence a detachment arrived at Isley Field, Saipan, on 15 December to provide coverage of the Twentieth Air Force bases on Saipan and Guam. At the end of February 1945, the squadron's ground echelon arrived at Iwo Jima where the air echelon joined them to become the first night fighter squadron based there.

Many missions from Iwo Jima comprised long-distance oceanic patrols. Sometimes the 548th NFS escorted wayward B-29s to the island when low on fuel or damaged by flak. The P-61s would locate such stragglers using their radar and then guide them to Iwo Jima. Around this time the squadron adopted Bablin, a black goat, as the unit mascot. The men painted Bablin's horns red to match the colour of the P-61's spinners.

On 12 June the 548th NFS moved to Ie Shima where it saw out the war with penetration raids and weather reconnaissance missions to assist planning B-29 missions over Japan. It was in this timeframe that the squadron claimed its first two aerial victories on 21 June. This was followed by two more on the night of 14-15 August, the last two aerial victories claimed by the USAAF in WWII. Following the end of the war the ground echelon relocated to occupied Japan in September, whilst its P-61s were placed in storage at Clark Field in the Philippines and Okinawa. The squadron was inactivated in December 1945.

The 548th NFS logo was in use for several months before it was officially approved on 7 August 1944. It comprises a fierce black tomcat wearing white gloves and holding a grey lit flashlight in its left paw. In its right paw it carries a revolver emitting smoke from the barrel, all the while striding across a fence against the backdrop of a large full moon. The logo is illustrated at the top of page 60.

The artwork on P-61A 42-5611 Midnite Miss, as shown in Profile 55, in early 1945.

Profile 51   P-61B serial 42-39525 *Night Take-Off*

This P-61B was among the second major shipment of replacement P-61Bs sent to the Pacific in late January 1945. Most of these were assigned to the 549th NFS, however, *Night Take-Off* joined the 548th NFS as a replacement aircraft.

Profile 52   P-61B serial 42-39375 *Hangar Lil*

This Black Widow was delivered to the 548th NFS on 11 October 1944 in Hawaii and was assigned to pilot First Lieutenant Robert Bertram and radar operator Second Lieutenant George Fairweather. Crew Chief Staff Sergeant Richard Baugh's biggest job was repairing *Hangar Lil* following a landing accident at Palmyra on 14 February 1945. No crew were injured, and the fighter was quickly repaired. On 24 July 1945 Bertram intercepted and shot down a Betty, previously shot up by a Hellcat. The aircraft was decommissioned when in storage at Clark Field on 5 August 1945.

A tired looking P-61B 42-39454 Cooper's Snooper, as depicted in Profile 57, seen towards the end of the war.

The imposing nose-art of P-61B 42-39405 Anonymous III, as shown in Profile 53.

Profile 53   P-61B serial 42-39405 *Anonymous III*

This Black Widow's assigned crew was pilot First Lieutenant Melvin Bode, radar operator Second Lieutenant Avery Miller and gunner Staff Sergeant John Hope. This was the third P-61 assigned to this crew following the crash landing of *The Spook* (see Profile 54 below), and the loss of serial 42-39413 *Spook II* when it was ditched after being hit by flak. The crew named *Anonymous III* when other 548th NFS crews refused to fly in another P-61 named *Spook* due to their suspicion that the name attracted bad luck.

Profile 54   P-61B serial 42-39403 *The Spook*

On 20 April 1945 the same assigned crew as noted for Profile 53 (above) were on an early morning patrol from Iwo Jima when they and three other P-61s became caught in a dense fog. P-61A serial 42-5610 *Midnight Madness* flown by Captain James Bradford blew a main tyre on the runway when he found the field in the fog. Following was *The Spook* which bounced off the top of *Midnight Madness* then crashed with its gear retracted. After it was scrapped for parts, the crew was assigned P-61B serial 42-39405 *Anonymous III* as a replacement.

Profile 55   P-61A serial 42-5611 *Midnite Miss*

This P-61A was included in the first batch of deliveries and was received by the 548th NFS in Hawaii on 21 September 1944. It was scrapped at Clark Field on 5 August 1945.

Profile 56   P-61B serial 42-39408 *Lady in the Dark*

On the night of 14-15 August 1945 this P-61 was unofficially credited with the last US aerial victory of WWII when crewed by pilot First Lieutenant Robert Clyde and radar operator Second Lieutenant Bruce LeFord. The pair claimed a Ki-44 Tojo without a shot being fired. When the Japanese pilot sighted the closing P-61, the Tojo descended to wave-top level and began a series of violent evasive manoeuvres, during which the fighter struck the water and shattered.

Profile 57   P-61B serial 42-39454 *Cooper's Snooper*

On 1 December 1944 this Black Widow was assigned to the crew of pilot First Lieutenant George Cooper, radar operator Second Lieutenant Stanley Babst and gunner Staff Sergeant Roy Ross. The artwork was painted by 548th NFS artist Staff Sergeant William Day. *Cooper's Snooper* was sent to Clark Field for storage on 1 December 1945 but was retrieved and redesignated as an F-61B on 1 July 1948. It was finally scrapped on 6 June 1949.

The commander of the 549th NFS, Lieutenant Colonel Joseph Payne (left), with Major Robert "Dave" Curtis, his 548th NFS counterpart. The pair are in front of P-61B serial 42-39430 Little Joe that was often flown by Payne. The fighter is depicted in Profile 61.

Pilot First Lieutenant Earl Tigner poses for the camera from the cockpit of P-61B 42-39424 Trigger Happy, as depicted in Profile 58.

CHAPTER 9
549th Night Fighter Squadron

The 549th NFS was established on 1 April 1944 at Hammer Field, California, and was one of the last dedicated USAAF night fighter squadrons to deploy to the Central Pacific. It first trained at airfields located in the San Joaquin Valley with both the P-61 and P-70. It was assigned to the Seventh Air Force and prepared for overseas deployment in October 1944.

The 549th NFS deployed to East Field, Saipan, in late February 1945 to provide night interceptor coverage for both Saipan and Guam. Saipan was a critical base at this juncture because long-range P-51 Mustangs were using it from which to escort B-29 bombers *en route* to Tokyo. A month later the squadron transferred to the newly captured Central Field on Iwo Jima. Here the USAAF combined SCR–527 and SCR–270 radars for early warning acquisition and the AN/TPS–10 radar for ground control of interception. This system enabled warning of intruders from about 140 miles distance. The ground control callsign "Li'l Abner" would then vector P–61s to engage. Often the intruders would drop chaff at 30 miles, blocking the early warning radar. When within ten miles of Iwo Jima, the P-61s would break contact to allow anti-aircraft batteries to engage.

During the final weeks of the war, the 549th NFS specialised in night intruder attack missions on enemy bases, toting loads of four 500-pound bombs or two 1000-pound bombs. These proved effective also against shipping and barges plying Japan's coastal waters. The squadron was awarded only one kill: on the night of 24 June 1945 pilot Flight Officer Gendreau claimed a Betty. The squadron remained at Iwo Jima until the war ended, concluding with long-range patrols over eastern China to supplement local night interdiction missions. It was demobilised whilst at Iwo Jima in early 1946.

Approved on 17 July 1944, the official 549th NFS logo features a grey bat in flight emitting white flashes while grasping two machine guns. The logo is illustrated on page 66.

First Lieutenant Donald Weichlein sitting in P-61B 42-39555 Night Nurse, named after his wife Mary, a nurse. The fighter is the subject of Profile 59.

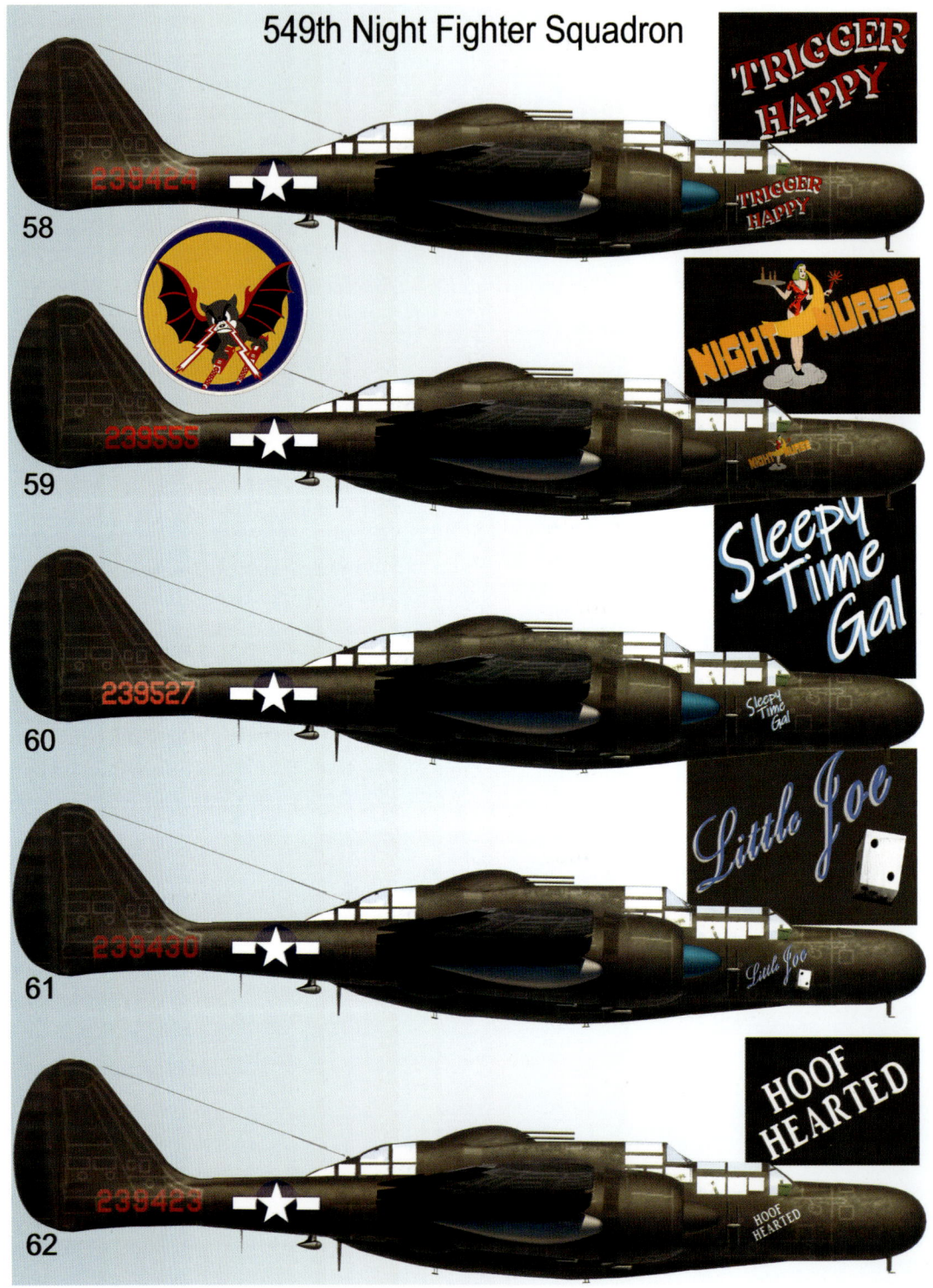

549th Night Fighter Squadron

Profile 58   P-61B serial 42-39424 *Trigger Happy*

This Black Widow was assigned to the 549th NFS on 27 January 1945. In May 1945 the fighter experienced a landing accident at a fog-bound Iwo Jima runway which damaged both maingears. Following repairs, the P-61B returned to service. It was ferried to Clark Field for storage on 5 February 1946, where it was scrapped there two years later.

Profile 59   P-61B serial 42-39555 *Night Nurse*

Arriving on the 549th NFS flightline on 28 January 1945, this P-61B's assigned aircrew was pilot First Lieutenant Donald Weichlein, radar operator Second Lieutenant Edward Mulvaney and gunner Sergeant William Dare. Weichlein named the aircraft to honour his wife Mary, a nurse. On 5 March 1945 during take-off the fighter climbed just above the runway and was in the process of retracting its undercarriage when the starboard engine failed and it crashed. Weichlein was injured in the back by a shattered propeller which entered the cockpit. The fighter was subsequently salvaged for parts.

Profile 60   P-61B serial 42-39527 *Sleepy Time Gal*

Another fighter carrying the most ubiquitous name for any USAAF WWII airframe, this P-61 was assigned into the 549th NFS from the Hawaiian Air Depot on 26 January 1945. The crew of pilot Major James Alford, radar operator Second Lieutenant Pat Condren and gunner Sergeant SI Coker were operating it on 23 May 1945 when Iwo Jima became covered in thick fog. A hard landing broke both maingears and damaged the radar equipment as well. Following repairs, on 2 November 1945 the aircraft was involved in a second landing accident, this time returning from a formation training flight. The aircraft was once again repaired, but it was soon ferried to Clark Field for storage. It was scrapped there in 1948.

Profile 61   P-61B serial 42-39430 *Little Joe*

This Black Widow was assigned to the 549th NFS on 27 January 1945 and was often operated by squadron commander Lieutenant Colonel Joseph Payne, who named the fighter. It was scrapped at Clark Field in 1948.

Profile 62   P-61B serial 42-39423 *Hoof Hearted*

This fighter only entered service after the end of the war in September 1945. It was briefly operated by pilot Second Lieutenant Robert Thum and radar operator Second Lieutenant Norman Drake. The name was a colloquialism of the times meaning mean-spirited. The airframe was soon condemned for salvage.

The crew of P-61B 42-39575 *Ohlami*, named after the three states of Ohio, Louisiana and Michigan. This fighter is the subject of Profile 68.

CHAPTER 10
550th Night Fighter Squadron

The 550th NFS was constituted on 3 May 1944 under the command of Major Robert Tyler. It first trained on the BT-13 Valiant and AT-10 Wichita before it finally received P-70s and P-61s. On 25 August 1944 the unit relocated to Visalia Army Air Field, California, where aircrews trained in specialised in night operations with these two models of night fighter. Reassigned to the Thirteenth Air Force, on 14 December 1944 the squadron arrived at Hollandia in Dutch New Guinea. From 31 December to 14 February 1945 a detachment operated from Middelburg Island.

On 14 February 1945 the 550th NFS moved to Morotai, then on 8 March it operated another detachment from Tacloban in the Philippines. A month later the remainder of the squadron moved to Tacloban, then from 27 April to 17 June 1945 it deployed another detachment from Zamboanga. In the final months of the war, it operated from Sanga Sanga on Sulu and Puerto Princesa on Palawan.

The original 550th NFS logo concept was of a goose firing two machine guns. This was replaced in the combat zone with the simpler logo of a blue disk with a full moon and two stars. This later logo was sometimes applied to aircraft tails. Both logos are illustrated on page 70.

P-61B 43-8238 Vivacious Vivian, as illustrated in Profile 69, showcases its new squadron number 332. In the background the nose artwork on P-61B 42-39473 Night Vision is visible (see Profile 67).

Profile 63   P-61B serial 42-39584 *The Prince*

This "B" model without a turret was delivered to the 550th NFS on 5 July 1945 from the Biak depot. It was then assigned to First Lieutenant Russel Lortz and radar operator Second Lieutenant Bob Boykin. Lortz' logbook shows *The Prince* flew 48 combat missions including several of eight hours duration to attack the oil refineries on Borneo. Note the gothic style name included narrow red piping within the letters. The fighter was placed in storage at Clark Field on 4 December 1945. In July 1948 it was converted to F-61B status but was decommissioned a few months later.

Profile 64   P-38L serial 44-24004

This P-38L served the 550th NFS in late 1944. Other P-38Ls known to have served with the squadron include serials 44-25559 and 44-26166.

The crew of P-61B 42-39485 The Gay-Blade, as depicted in Profile 65. Pilot Major Victor Modina is wearing a sombrero to celebrate his Mexican heritage.

The crew of P-61B 42-39473 Night Vision pose with their aircraft, which is illustrated in Profile 67.

PACIFIC PROFILES

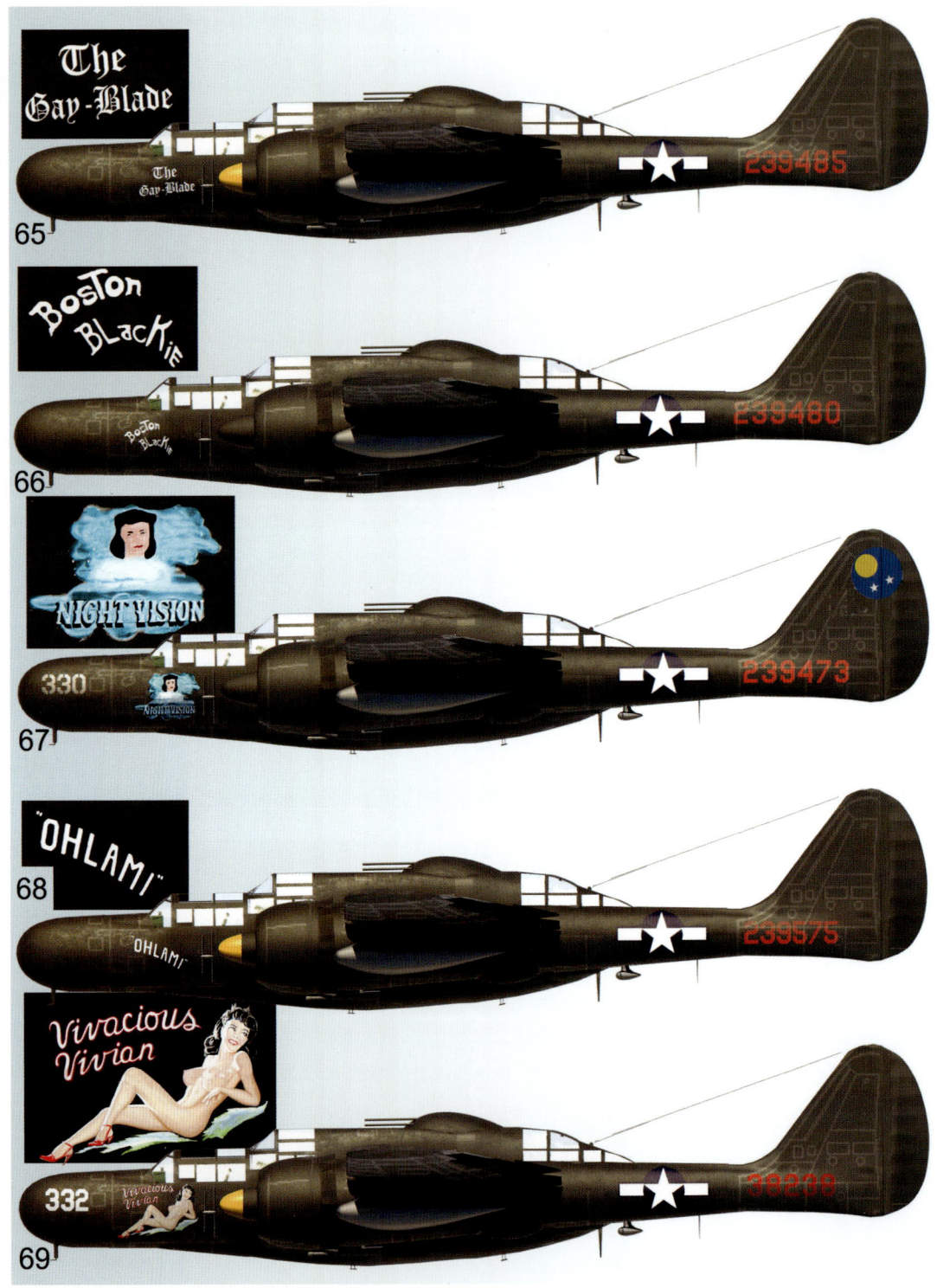

Profile 65   P-61B serial 42-39485 *The Gay-Blade*

This Black Widow was assigned to the 550th NFS from the Biak depot on 5 February 1945. The name *The Gay-Blade* references a knife that pilot Major Victor Modina carried on every mission. After the war this fighter was stripped of all armour-plating and radar equipment so it could be used as a trainer primarily for pilots to maintain flying hours. It was decommissioned in 1947.

Profile 66   P-61B serial 42-39480 *Boston Blackie*

This Black Widow was named by assigned pilot Captain Richard Meldon after a character in 1940s radio and films, Boston Blackie, a reformed jewel thief who became a detective. It was received into the 550th NFS on 1 February 1945 from the Biak depot. First Lieutenant Melvin Stevick was conducting a strafing mission on 5 April 1945 when a shell exploded near the port maingear, damaging the tyre. The aircraft crashed back at base but was repaired back to airworthy status. It was converted to an F-61B in 1948 and finally decommissioned in 1949.

Profile 67   P-61B serial 42-39473, squadron number 330, *Night Vision*

When the 419th and 550th NFS were collocated at Puerto Princesa, this Black Widow was transferred to the latter squadron. It had originally been named *Saddle Happy* when with the 419th NFS (see Profile 31) and was renamed *Night Vision* after night-vision binoculars were first fitted to "B" models to better see targets. An APS-13 tail-warning radar was often added at the same time. This fighter was finally decommissioned in 1949.

Profile 68   P-61B serial 42-39575 *Ohlami*

This P-61B was first assigned to the 550th NFS on 28 February 1945, and was operated by pilot First Lieutenant Newell Witte, radar operator Second Lieutenant Henry O'Brien and gunner Staff Sergeant Tom Cope. The three men created the name *Ohlami* by combining the two-letter codes of their home states: Ohio, Louisiana and Michigan. The fighter was transferred to the 418th NFS in Japan just after the war ended but was destroyed during a crash landing on 14 April 1946 at Kyushu.

Profile 69   P-61B serial 43-8238, squadron number 332, *Vivacious Vivian*

This late model "B" Black Widow was shipped to the Pacific on 30 May 1945 where it was assigned to the 550th NFS at Puerto Princesa. At this late stage of the war the squadron had started assigning squadron numbers in the 300 series, with *Vivacious Vivian* assigned number 332. The fighter was finally scrapped in 1948.

Three VF(N)-101 Corsairs, including #13 and #14, aboard the carrier USS Intrepid. Note the reddish brown etch primer application to the forward radar domes.

CHAPTER 11
VF(N)-75 & VF(N)-101

In January 1943 VF(N)-75 became the first USN Corsair night fighter squadron established in the US. A detachment was deployed to the Pacific in October 1943 with six F4U-2s, commensurate with VMF(N)-531's PV-1 Ventura deployment (see Chapter 12). Commanded by Lieutenant Commander William "Gus" Widhelm, VF(N)-75 arrived in the theatre with eighteen pilots. Only five of these had instrument flying experience, three had Corsair time, three had F4F Wildcat experience and the balance had only flown the basic SNJ trainer.

The VF(N)-75 detachment arrived at Guadalcanal to deter the serial pest "Washing Machine Charlie", in fact numerous and different types of Japanese aircraft sent over at irregular intervals as nuisance raids. Shortly thereafter the squadron moved to Munda, from where its first patrol was on 2 October 1943. The unit's first success was on 31 October when Lieutenant Hugh O'Neil shot down Warrant Officer Shinohara Hisami's No. 751 *Ku* Betty off Vella Lavella at 2115. From Munda the squadron moved successively to Torokina, Green Island and finally Emirau. Land-based combat operations for the VF(N)-75 detachment ended in early February 1944.

Meanwhile in December 1943 VF(N)-75 had received ten F4U-2s and was ordered to Hawaii for carrier deployment. These were initially designated detachments Ten and Eleven and were assigned to the carriers USS *Enterprise* and USS *Intrepid* respectively. The two detachments were restructured as VF(N)-101 on 1 January 1944 under the command of Lieutenant Commander Richard "Chick" Harmer, who had served as VF(N)-75's executive officer. The newly named squadron spent the first two weeks of January practicing night flying from Barbers Point Naval Air Station.

On 15 January, VF(N)-101 became the USN's first carrier-based night fighter squadron when Harmer took a detachment of four Corsairs aboard USS *Enterprise* (squadron numbers 9, 10, 11 and 15). The next day, *Enterprise* joined Task Force 58.1 for a six-month deployment to strike Truk, Palau, Marshall Island and the Mariana Islands. Hugh Gray, a civilian engineer from Sperry, was aboard the voyage as an advisor. Meanwhile a second detachment of four Corsairs, led by Lieutenant Cecil "Swede" Kullberg, boarded the carrier USS *Intrepid*.

VF(N)-101's first engagement was in the early hours of 19 February 1944 following which Commander Harmer was credited with damaging a G4M (Betty) northeast of Pohnpei, in the Caroline Islands. The squadron's final combat was on 28 June 1944 when Harmer was credited with shooting down another Betty.

USS *Intrepid*, meanwhile, was unable to conduct flight operations following steering problems from an enemy air attack, and she made her way back to Pearl Harbor in March 1944 and then to San Francisco for major repairs. At Pearl the VF(N)-101 detachment was put ashore and returned to Barbers Point. There the squadron was made an experimental and training squadron with Captain Jack Griffin appointed as the new commander.

To minimise visibility, neither carrier-based detachments of VF(N)-101 nor the land-based VF(N)-75 detachment applied squadron numbers to the sides of the fuselage. Instead, both stencilled the numbers on the forward-facing undercarriage doors, and VF(N)-75 also applied small-sized numbers on both sides of the forward cowl. The carrier-based Corsairs also had the forward section of their radar domes painted with a dull matt reddish brown etch primer to prevent sea corrosion.

Lieutenant Commander Richard "Chick" Harmer flies his VF(N)-101 F4U-2 number 15 from the USS Enterprise somewhere over the central Pacific in mid-1944.

Profile 70   F4U-2 VF(N)-75 squadron number 5

The VF(N)-75 Corsair inventory appeared in the USN two-tone camouflage scheme. This F4U-2 is illustrated as it appeared at Torokina in January 1944.

Profile 71   F4U-2 VF(N)-101 squadron number 15

Lieutenant Commander Richard "Chick" Harmer flew this Corsair from the USS *Enterprise* during the first half of 1944 in the central Pacific.

Profile 72   F4U-2 VF(N)-101, Bu Aer 02710, squadron number 10

Lieutenant Bob Brunson crashed this Corsair when landing aboard USS *Enterprise* at night. He managed to walk away from the accident uninjured although the fighter was a write-off.

PV-1 Ventura 29811 Eight Ball, as depicted in Profile 74, undergoing routine maintenance at Torokina on 25 June 1944.

CHAPTER 12
VMF(N)-531

VMF(N)-531 was commissioned on 16 November 1942 in North Carolina and then began training as the first Marine Corps night fighter unit. The unit underwent many challenges. In February 1943 the squadron sent nine pilots to England to learn RAF night fighter tactics. These men were later instructed on the SCR-527 early warning radar, which the squadron initially took into combat. Prior to this squadron commander Lieutenant Colonel Frank Schwable had learned about RAF night fighter operations against the Luftwaffe in 1941.

Subsequently the USMC received authorisation to establish eight night fighter squadrons in the first half of 1943, named Project Affirm. The Guadalcanal campaign rearranged priorities with the urgent creation of VMF(N)-531 in November 1942. However, after a lengthy training period the squadron did not report to COMAIRSOLS for duty until August 1943.

The first VMF(N)-531 detachment comprising six Venturas was allocated squadron numbers 1 to 6. These departed Hawaii on 19 August 1943 and arrived at Espiritu Santo six days later. The contingent then proceeded to the Russell Islands the following month and commenced air patrols. On 23 October Ventura #5 undertook a night bombing mission over Buin with 100-pound bombs. Such missions were encouraged to enable pilots to develop theatre experience until intercept missions could be flown. Due to strict security around night operations, other units in the theatre largely remained unaware of the night fighters' existence. By late October the squadron was cooperating with its own Ground Control Intercept personnel, located on remote Liapari Island, to the south of Vela Lavella. The squadron eventually operated from the Russell Islands, Vela Lavella and Bougainville all the while pioneering USMC night fighter tactics.

On 16 September VMF(N)-531 lost its first Ventura, #51 *Coral Princess*, when it failed to return from a practice intercept mission over the Russell Islands. Then in the early morning of 31 October a Corsair lost control on take-off, breaking PV-1 #6 *Miss Represented* in half.

On 13 November 1943 VMF(N)-531 claimed its first night kill, while under the control of "Horse Base", a task force sailing around 50 southwest of Torokina. The victim was a No. 751 *Ku* G4M1 Betty. On 3 December PV-1 #54 went missing during a night mission from Barakoma. Ground radar tracked it pursuing an enemy aircraft over USN destroyers approaching Bougainville, and surmised it had collided with its quarry. Three nights later the squadron's first collaborative GCI radar/night fighter team kill was made by Lieutenant Colonel John "Iron John" Harshberger, who downed a Jake over Empress Augusta Bay on Bougainville with a combination of forward and turret gunfire.

At this juncture a spate of losses commenced. On 9 February 1944 PV-1 #57 crashed into Vela Gulf just after take-off, then ten days later PV-1 #56 drove into the ocean whilst searching for barges south of the Green Islands. By the end of February VMF(N)-531 was down to ten Venturas. Then on 13 March PV-1 #54 crashed on a night take-off then PV-1 #58 was written off the following day during a runway collision with a Corsair. On 21 March #50 and #69 collided during a formation switchover, costing the lives of two four-man crews.

Rabaul's air power was withdrawn to Truk at the end of February 1944 at a time when high demand was taking heavy toll of the Venturas. On 7 April at Green Island PV-1 #53's maingear collapsed during landing and soon thereafter it along with #51, #52 and #62 were condemned as unairworthy. Then #55 also suffered landing gear failure during landing, becoming another write-off. Some missions persisted as far as Rabaul and on 9 May an attempted interception of a Jake in that area failed when the floatplane, alerted by the gun flashes, dove away.

The fast Ventura was often difficult to slow down sufficiently to engage enemy aircraft, particularly slower floatplane types. On 12 January Commander Schwable claimed a B5N Kate offshore Torokina Point. Then on 5 February he claimed the fifth victory for the squadron against a Betty in the same area, in fact a No. 751 *Ku* Betty commanded by FPO1c Itou Takashi.

Although combat patrols were sometimes flown with two pilots, standard operating procedure was to fly night missions with one pilot, a turret gunner and a radar operator. Schwable regarded the PV-1 as only an interim solution. He reported that the instruments were difficult to read at night, and the airframe was equipped with unreliable armament and radar. With a service ceiling lower than operational expectations, Schwable suggested *inter alia* removing the Martin turret and upper guns as a way of improving the type's performance.

Hence VMF(N)-531 subsequently undertook field modifications that included stripping out all unnecessary gear, such as flares and bomb bay doors. Some PV-1s also removed their turrets. The combined effect of these removals improved the type's rate of climb but not its speed. The remaining and persistent problem was interference to the radar systems from high voltage ignition from the engines. The type still climbed too slowly, had limited altitude and was considered a poor instrument platform. By early 1944 it was apparent that additional PV-1s would be not allocated given their diminishing role in the Pacific. In May 1944 COMSOPAC requested the unit be withdrawn to the US for re-equipping, however, Washington took several months to mull the decision.

On 15 May 1944 VMF(N)-531 moved operations to Torokina on Bougainville, before ending its South Pacific tour from Green Island conducting night CAPs for PT boats and other small USN vessels. The squadron wound up Pacific operations on 14 July 1944 with its last mission a dusk patrol over Rabaul. It claimed a total of twelve victories by five different crews, all at night.

VMF(N)-531's five remaining airworthy PV-1s returned to Hawaii on 11 August. By now the unit had earned the sobriquet "The Grey Ghosts". The unit was reactivated on 13 October 1944 at Eagle Mountain Lake near Fort Worth, Texas, and in 1945 converted to the twin-engine Grumman F7F Tigercat. The experience of operating the PV-1 contributed to developing subsequent USMC night fighter doctrine.

The VMF(N)-531 logo was designed by original squadron member Lieutenant Colonel Jack Colby in the US before its overseas deployment. Colby designed a shrouded skeleton depicting the Grim Reaper reaching out to touch its next victim, against a black background with a thin crescent moon. The stars are grouped into clusters of five, three and one, reflecting squadron designation 531, and are arranged according to the constellation Orion which represents a warrior. Lightning bolts coming from the eyes represent the squadron's use of radar to detect the enemy. Colby's striking motif continued with post-war use, with two versions illustrated at the top of page 82.

VMF(N)-531's initial assigned fifteen PV-1s were gathered from other training units and had little conformity in markings. They had a variety of USMC insignia, varying from a single star to the star-and-bar including some with the red surround from the August 1943 specification. These insignia were modified in the field: the single stars had white bars added whilst some red surrounds were painted over in black or dark blue (see Profile 76).

The USMC was less strict than its USN counterparts in personalising aircraft, with some commanders allowed modest nose-art including unit commander Schwable who permitted naming and decoration of airframes. The paint on these was badly weathered in the tropics. Anti-icing leading edges were retained with the anticipation aircraft would encounter icing conditions on night operations, and this was the case.

The advance detachment of six Venturas were numbered one to six, however, shortly after arrival in the theatre they were renumbered to conform to fit in with the extant MAG-21 two-digit allocation system. Squadron numbers known to have been allocated (aside from 1 to 6) include 25, 30, 46, 51, 52, 53, 54, 55, 56, 57, 58, 62 and 69. Most were applied in a large black stencil on the cowl although at least one (see Profile 78) had the stencil applied in white. VMF(N)-531 lost eleven airframes to accidents during its time in the South Pacific.

PV-1 Ventura Gertie the Goon at Banika Field in the Russell Islands shortly after its arrival from Hawaii, and before it was over-sprayed with matt black paint. Note the two-tone camouflage drop tank. The aircraft is the subject of Profile 78.

Profile 73   PV-1 Bu Aer unknown

Lieutenant Colonel Jack Colby's VMF(N)-531 logo was applied to this airframe just prior to the unit deploying overseas. The shrouded Grim Reaper was applied over a square matt black background just forward of a star-and-bars insignia. The aircraft is featured as it appeared in the US, with its unknown squadron number yet to be applied.

Profile 74   PV-1 Bu Aer 29811 *Eight Ball*

This Ventura arrived in the Russell Islands on 19 October 1943 as a replacement aircraft. Its assigned crew named their aircraft *Eight Ball*, a slang term for habitual slackers of the time. They applied the name and motif to the nose as well as painting both maingear hubs with the same motif. They also memorialised the previous Jenkins crew, lost on the night of 3 December 1943, by stencilling their names as illustrated. This Ventura was from the second production run of PV-1s and has a more pointed nose than the first but retained the nose windows. It was among the squadron's five remaining airworthy PV-1s which returned to Hawaii in August 1944. For some reason it had no squadron number applied and was delivered in the USN three-tone camouflage scheme with the red-surround star-and-bar insignia.

Ventura 29854 Chloe, as depicted in Profile 77, at Vella Lavella on 13 January 1944.

VMF(N)-531 personnel pose in front of a Ventura prior to departing from the US for South Pacific service. The aircraft features Lieutenant Colonel Jack Colby's VMF(N)-531 logo, as is shown in Profile 73.

PACIFIC PROFILES

Profile 75   PV-1 Bu Aer 29755, squadron number 51, *Coral Princess*

At around 2000 on 16 September 1943 this Ventura went missing when it failed to return from a practice intercept mission over the Russell Islands. The pilot last radioed that he was south of Savo Island. The nose art features a Vargas calendar girl

Profile 76   PV-1 Bu Aer unknown, squadron number 3

This PV-1 was one of the six in the first VMF(N)-531 detachment numbered #1 to #6. After departing Hawaii these arrived at Espiritu Santo on 25 August. The red surround on the star-and-bar was painted over in black shortly after arrival in theatre, as illustrated.

Profile 77   PV-1 Bu Aer 29854, squadron number 53, *Chloe*

This bomber was named in the US and was one of the Venturas which had its Martin turret removed to improve its rate of climb. When landing on 7 April 1944 at Green Island its maingear collapsed and alongside #51, #52 and #62 it was decommissioned as unairworthy.

Profile 78   PV-1 Bu Aer unknown, squadron number 52, *Gertie the Goon*

This Ventura had its maingear wheel hubs decorated as illustrated, and following considerable combat it was decommissioned at Green Island in April 1944 as unairworthy. The airframe was sprayed overall matt black shortly after it arrived in the theatre with the squadron number stencilled in white. The black paint quickly wore through in the harsh climate, however, giving a hybrid appearance.

Profile 79   PV-1 Bu Aer 29756, squadron number 46

This Ventura was the first of the initial batch of fifteen received in the US on 15 February 1943. Transferred from a training unit it appeared in overall matt black with a single star insignia. The aircraft is illustrated after it was renumbered #46 and had white bars added to the star insignia after arrival in the South Pacific.

F4U-2 #201 Shirley June, as flown by VMF(N)-532 commander Major Everett Vaughn in 1944. It is depicted in Profile 80.

CHAPTER 13
VMF(N)-532

The founding commander of VMF(N)-532 was Major Ross Mickey who in 1943 was allocated eighteen of the three dozen F4U-2 Corsairs built. These were assigned squadron numbers 201 to 218. He was followed by Major Everett Vaughan, who developed USMC tactics and methods to take advantage of the radar-equipped Corsair. A previous airline pilot, Vaughan arrived at Cherry Point in the spring of 1943. The squadron deployed to Tarawa on 13 January 1944, operating under the umbrella of MAG-31. Working with ground controllers the Corsairs undertook intercepts but never made contact with the enemy.

On 15 February 1944 VMF(N)-532 moved to Kwajalein in the Marshall Islands where Major Everett Vaughn flying his #201 *Shirley June* was the first to land on recaptured Japanese mandated territory. At Kwajalein the squadron later linked up with Air Warning Squadron One (AWS-1) which had advanced ground control radar to track night raiders which were proving a major problem at Roi.

From that location on 14 April 1944, Lieutenant Edward Sovik and Captain Howard Bollman each claimed a Betty bomber off Eniwetok atoll, VMF(N)-532's only kills of the war. The Corsairs also conducted night raids against Wotje in the Marshall Islands and later night patrols from Saipan and Guam. The squadron returned to the US in October 1944.

The VMF(N)-532 logo featured a diving black panther silhouetted against a yellow moon, surrounded by white stars. It is illustrated at the top of page 88.

F4U-2 Shirley June, the subject of Profile 80, being refuelled at Roi in the Marshall Islands.

PACIFIC PROFILES

VMF(N)-532

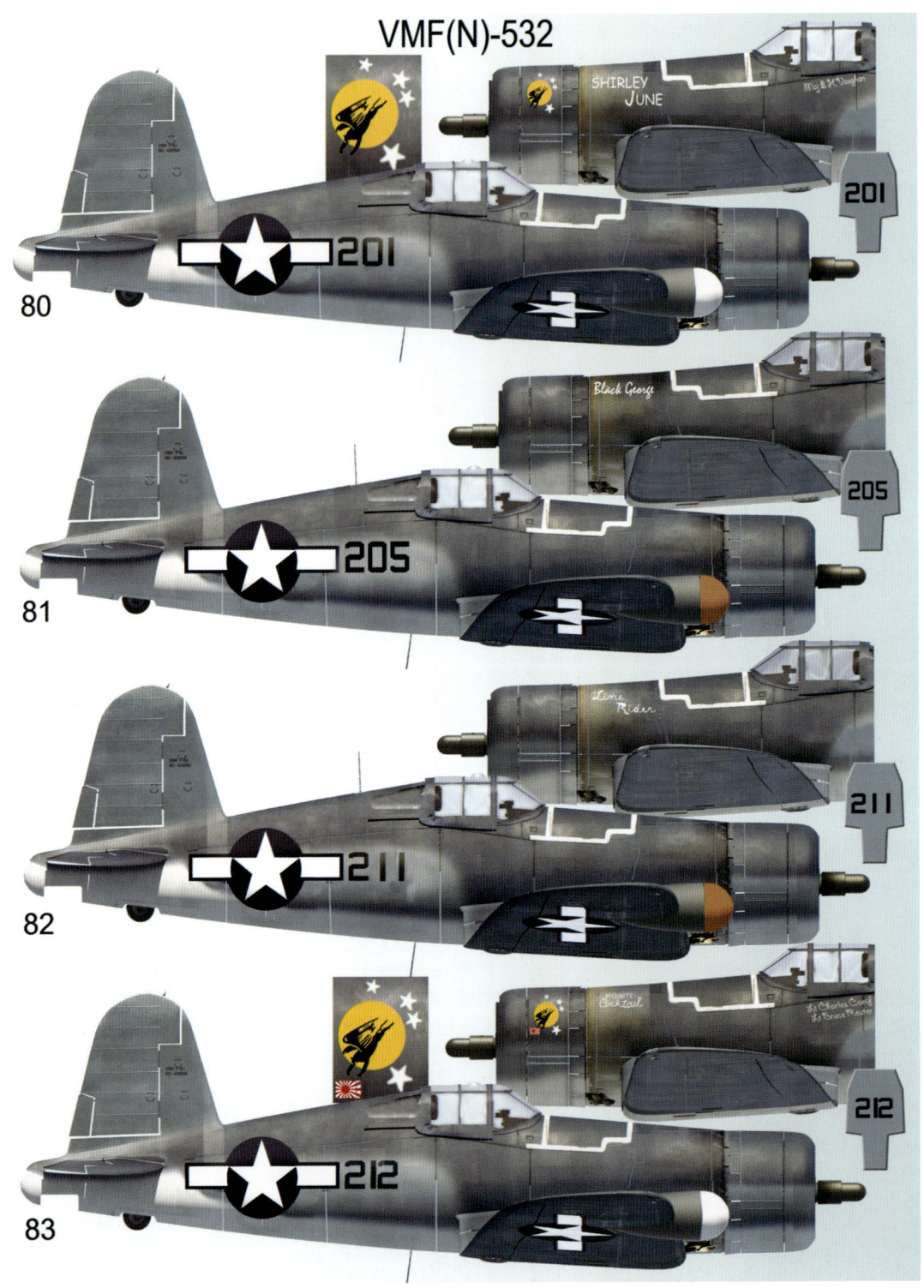

Profile 80   F4U-2 squadron number 201 *Shirley June*

Assigned to Major Everett Vaughn, *Shirley June* was the first American aircraft to land on captured Japanese mandated territory in February 1944.

Profile 81   F4U-2 squadron number 205 *Black George*

Assigned to Lieutenants Paul Dolhude and Edward Sovik, Captain George Hubbard also flew this fighter regularly. It is illustrated as it appeared in June 1944 at Saipan.

Profile 82   F4U-2 squadron number 211 *Line Rider*

This VMF(N)-532 Corsair is illustrated as it appeared in June 1944 at Saipan.

Profile 83   F4U-2 squadron number 212 *Midnite Cocktail*

This VMF(N)-532 Corsair was being flown by Captain Howard Bollman when he shot down a Betty on 14 April 1944. The Japanese victory flag just beneath the squadron motif represents Bollman's first claim, a Zero while he was assigned to VMF-112 in the Solomons.

Captain George Hubbard astride F4U-2 205 Black George, as illustrated in Profile 81. Note the unconventional combat footwear.

P-38H 42-66684 that was used to tow aerial targets for the 419th NFS, as depicted in Profile 86. It is seen at Fighter #2 on Guadalcanal before it had its serial number reapplied on the fin.

Brigadier General Earl Barnes arrives at Morotai in late 1944 in P-61A-5-NO 42-5531 (see Profile 85). The navigator/radio operator is opening the hatch as Barnes shuts the engines down.

CHAPTER 14
Miscellaneous

This chapter presents five profiles of aircraft directly associated with Pacific night fighters although they were not specifically assigned to the role.

Profile 84   P-39D serial 41-38401, 35th FS, squadron letter M

The overall matt black scheme of this 35th FS Airacobra is due to a night fighter experiment tried following the July 1942 night attacks on Townsville by No. 14 *Ku* H8K Emily flying boats. Squadron commander Major Norman "Coach" Morris identified his fighter with the letter "M", which was also the first letter of not only his surname, but also his rank and coincidentally the surname of his crew chief, Staff Sergeant Matteo.

Profile 85   P-61A-5-NO serial 42-5531

This Black Widow was first assigned into the 419th NFS at Guadalcanal in May 1944. In late 1944 it was retired from combat and personalised for the use of Brigadier General Earl Barnes, the commander of Thirteenth Fighter Command. The fighter sported the 550th NFS emblem on the tail as it was serviced by that unit. The airframe was stripped back to natural metal finish, and the radar was removed with the fibreglass radome being replaced by aluminium. A radio direction finder was fitted, and the turret was replaced with a B-24J fuel tank to enhance range.

Profile 86   P-38H 42-66684, 18th FG

Following service with the 339th FS, this P-38H had all armament removed and was painted overall yellow to make it highly visible. It was then used to tow gunnery targets at Guadalcanal for the P-61s of the 419th NFS, as well as the P-38Js of the 44th and 70th FS.

Profile 87   P-40N serial unknown, *Rusty*, 418th NFS

Shortly after the 418th NFS took delivery of its P-61s at Nadzab, the squadron's pilots were given courses in fighter tactics which centered on operating the Black Widows in pairs for defensive purposes. Squadron commander Major William Sellers subsequently acquired this P-40N for training purposes which he named *Rusty* after his young son. He took glee in jumping the new P-61As during daytime training flights to ascertain mistakes the new P-61 pilots made under pressure.

Major Norman "Coach" Morris at Milne Bay with P-39D 41-38401, as illustrated in Profile 84.

Profile 88   Beaufighter Mk 1c, A19-28, squadron letter P, No. 30 Squadron, RAAF

Among the most unusual makeshift night fighter in the Pacific was the RAAF Beaufighter, which otherwise had a niche in conducting low-level attacks against Japanese shipping and supplies. Following a nocturnal raid on Port Moresby by four No. 751 *Ku* Bettys on 22 October 1943, No. 30 Squadron anticipated a similar follow-on raid and began sending up Beaufighters to conduct night patrols in pairs.

Two nights later the Bettys returned, but the RAAF pilots were unable to make a successful interception. The No. 30 squadron diary recorded:

> There was a lot of confusion upstairs last night. The Japs sailed serenely over the target and straight out again while our boys covered miles searching for them. Parker caught a glimpse of one and swung in behind, but the Jap plane disappeared into thick cloud at that moment.

This was Squadron Leader Peter Parker in A19-28 who recorded seeing a solitary enemy bomber. Photos of A19-28 at this juncture shows its lower fuselage half had been repainted, however, it is doubtful this was a matt black. Instead, the original RAF camouflage scheme as applied in the UK wore off quickly in the tropics and the paint was more likely dark RAAF Foliage Green.

Seen with another Beaufighter over Port Moresby in late 1943, A28-18 "P" (lower) is the subject of Profile 88.

Sources & Acknowledgments

This volume draws only on primary sources. The author's extensive collection of documents, diary extracts, photos and notes from field trips contributed a plethora of information, too extensive to list. Mainstream sources include, *inter alia*:

Headquarters Fifth Air Force Special Orders No. 272, 29 September 1943

Allied Translator and Interpreter Section (ATIS) Reports

Allied Air Force Intelligence Summaries (AWM)

Flightpath article "Kept in the Dark", third quarter 2010

1962/63 ANGAU patrol report by J. G. O'Brien.

Official USAAF histories (microfilm) for 6th, 418th, 421st, 547th, 548th, 549th, 550th and 419th NFS.

Japanese combat logs for 702nd and 705th *Kokutai* (*kodochosho*)

Keith Rundle RAAF patrol reports 1962 Bogia/Madang region

Diary, John Florence of 6th NFS

Fifth Air Force, correspondence and technical reports 1943-45

CILHI site reports, numerous post-war

Port Moresby radar log 15 May 1943.

P-61 loss details contained in Individual Deceased Personnel Files (IDPF)

Field Trips by author throughout New Guinea and the Pacific, 1964-2017

Pacific Aircraft Historical Society - Wreck Data Sheets

PNG Colonial Office - Civil Administration Records and PNG Cultural Museum

British Solomons Colonial Office - Civil Administration Records

Aircraft Movement Entries, Townsville Control Tower, 1943-44

Memoirs, Northrop representative and test pilot John Meyers

Records, Fifth, Seventh and Thirteenth AF Units via Maxwell AFB

Records, Air Forces Establishment, Fifth Fighter Command, Thirteenth Fighter Command. Thirteenth Bomber Command, 4th Air Depot and 27th Air Depot.

Factory Deliveries and Departures from USA, By Type, Model and country, 1944/45 (AHRA, Maxwell AFB)

USAAF individual aircraft record cards (AFHSO, Bolling AFB).

War Department AN-01-60G B-25C and D Series (1943)

Directorate of Technical Services P-61 Technical Instruction Manual

Relevant USMC and USN squadron histories (NARA and database Fold3)

Website www.pacificwrecks.com and its diligent webmaster, Justin Taylan

Index of Names

Acre, Second Lieutenant John   28
Adams, Second Lieutenant Burrell   20
Alford, Major James   67
Allain, Second Lieutenant Byron   51
Anderson, Staff Sergeant William   27
Avery, Tex   53
Babst, Second Lieutenant Stanley   63
Barker, Major Emerson   39
Barker, First Lieutenant Max   37
Barnes, Brigadier General Earl   90, 91
Barnett, Lieutenant Ira   35
Baugh, Staff Sergeant Richard   60
Bennett, Captain Earl   20
Berg, Second Lieutenant Alexander   51
Bertram, First Lieutenant Robert   60
Black, Second Lieutenant George   43
Bode, First Lieutenant Melvin   63
Bollman, Captain Howard   87, 89
Bong, Lieutenant Richard   47
Bourque, First Lieutenant Arthur   56
Boykin, Second Lieutenant Bob   70
Bradford, Captain James   63
Bradford, Staff Sergeant Joseph   51
Bradley, Captain William   47, 48, 51
Braxton, Staff Sergeant Millard   51
Brill, Sergeant Leon   28
Brunson, Lieutenant Bob   77
Clancy, Sergeant Donald   44
Clyde, First Lieutenant Robert   63
Coker, Sergeant SI   67
Colby, Lieutenant Colonel Jack   80, 82, 83
Condren, Second Lieutenant Pat   67
Cooper, First Lieutenant George   63
Cope, Staff Sergeant Tom   73
Corts, First Lieutenant David   51
Craig, Lieutenant Edward   31, 34
Crough, Technical Sergeant John   54, 56
Curtis, Major Robert "Dave"   64
Daniel, Captain Howard   43, 44

Dare, Sergeant William   67
Day, Staff Sergeant William   63
Dolhude, Lieutenant Paul   89
Drake, Second Lieutenant Norman   67
Duethman, Second Lieutenant Raymond   35
Dutkanicz, Private Peter   27
Fairweather, Second Lieutenant George   60
Falcon, Sergeant Adolfo   43
Farrelly, Sergeant Patrick   27
Ferguson, Second Lieutenant Paul   43
Ferguson, Second Lieutenant Robert   28
Forrestal, James   33
Forrestal, Second Lieutenant Robert   33, 34
Freeman, Second Lieutenant George   53
Frenzel, First Lieutenant Robert   43
Fuller, Staff Sergeant Cecil   43
Furuya Sadao, Warrant Officer   20
Gendreau, Flight Officer   65
Gray, Hugh   75
Griffin, Captain Jack   75
Grotzinger, Second Lieutenant Tony   44
Haberman, First Lieutenant Dale "Happy"   27
Hans, First Lieutenant Phillip   44
Hansen, Second Lieutenant Jerome   27
Harmer, Lieutenant Commander Richard   75-77
Harshberger, Lieutenant Colonel John   79
Hinz, Flight Officer Daniel   27
Holiday, Lieutenant Hap   37
Holley, Flight Officer Harold   33
Holloway, Second Lieutenant "Doc"   44
Hope, Staff Sergeant John   63
Hornaday, Captain Warren   2, 19, 20, 23
Hubbard, Captain George   89
Itou Takashi, FPO1c   80
Johnson, Colonel Robert   54, 56
Kamajian, Second Lieutenant George   37
Kane, First Lieutenant Loren   44
Kaplan, Sergeant Rubin   53
King, Admiral Ernest   8

Kohl, First Lieutenant Fred 53
Kullberg, Lieutenant Cecil "Swede" 75
Kunzman, Second Lieutenant Robert 53
Kuzmick, Lieutenant Alexander 47
Laffey, Captain John 43
LeFord, Second Lieutenant Bruce 63
Lockard, First Lieutenant Al 53
Logan, First Lieutenant Stan 37
Lortz, First Lieutenant Russel 70
Mahr, Major Victor 25
Malone, First Lieutenant Tom 47, 53
Matteo, Staff Sergeant 91
McCloskey, Captain John 20
McCumber, Second Lieutenant Myrle 27
McIvor, First Lieutenant William 21
Meldon, Captain Richard 73
Meyer, Second Lieutenant John 19
Mickcy, Major Ross 87
Miller, Second Lieutenant Avery 63
Miozzi, Staff Sergeant Leroy 25, 28, 29
Modina, Major Victor 71, 73
Mooney, Second Lieutenant Raymond 20, 27
Morris, Major Norman "Coach" 91, 92
Mulvaney, Second Lieutenant Edward 67
Myers, John 11, 47
Nielsen, First Lieutenant Al 37
Nowak, Sergeant Walter 56
O'Brien, Second Lieutenant Henry 73
O'Neil, Lieutenant Hugh 75
Oates, First Lieutenant Robert 43
Olley, Lieutenant John 48
Owen, Second Lieutenant John 51
Ozmun, Second Lieutenant Harold 43
Parker, Squadron Leader Peter 93
Pavlecka, Vladimir 8
Payne, Lieutenant Colonel Joseph 64, 67
Pharr, Major Walter 47, 48
Piett, Second Lieutenant Homer 37
Porter, Second Lieutenant Philip 37
Ridenour, Sergeant MW 44
Ross, First Lieutenant William "Bill" 35
Ross, Staff Sergeant Roy 63

Rucks, Lieutenant Bonnie 54, 56
Savaria, First Lieutenant Robert 53
Scheiber, Captain Ken 54, 56
Schwable, Commander Frank 79-81
Sellers, Major William "Bill" 30, 31, 37, 92
Shinohara Hisami, Warrant Officer 75
Shozo Kohira, Warrant Officer 27
Smith, Major Carroll 31, 37
Smith, First Lieutenant Hoke 51
Sovik, Lieutenant Edward 87, 89
Stevick, First Lieutenant Melvin 73
Straub, Staff Sergeant Fred 43
Sumita Shinshichi, Lieutenant 20
Sutliff, Captain Donald 49
Taylan, Justin 47
Thomas, Captain Ernest 28
Thompson, First Lieutenant Warren 23
Thum, Second Lieutenant Robert 67
Tigner, First Lieutenant Earl 64
Trabing, Sergeant Donald 51
Tyler, Major Robert 69
Vanderhoff, Second Lieutenant Dean 2
Vaughn, Major Everett 86, 87, 89
Wallace, Second Lieutenant William 27
Ward, Second Lieutenant Charles 28
Weichlein, First Lieutenant Donald 65, 67
Widhelm, Lieutenant Commander William 75
Winslow, First Lieutenant Robert 44
Witte, First Lieutenant Newell 73
Wolf, First Lieutenant Owen 51
Wood, Second Lieutenant Richard 37
Woodring, Second Lieutenant Alton 51
Yahn, Sergeant William 43
Zimmer, Lieutenant Paul 47, 51